The Human Brain during the Second Trimester 190- to 210-mm Crown-Rump Lengths

This tenth of 15 short atlases reimagines the classic 5-volume *Atlas of Human Central Nervous System Development*. This volume presents serial sections from specimens between 190 mm and 210 mm with detailed annotations. An introduction summarizes human CNS developmental highlights around 5 months of gestation. The Glossary (available separately) gives definitions for all the terms used in this volume and all the others in the *Atlas*.

Key Features

- Classic anatomical atlas
- Detailed labeling of structures in the developing brain offers updated terminology and the identification of unique developmental features, such as germinal matrices of specific neuronal populations and migratory streams of young neurons
- Appeals to neuroanatomists, developmental biologists, and clinical practitioners
- A valuable reference work on brain development that will be relevant for decades

ATLAS OF
HUMAN CENTRAL NERVOUS SYSTEM DEVELOPMENT
Series

Volume 1: The Human Brain during the First Trimester 3.5- to 4.5-mm Crown-Rump Lengths

Volume 2: The Human Brain during the First Trimester 6.3- to 10.5-mm Crown-Rump Lengths

Volume 3: The Human Brain during the First Trimester 15- to 18-mm Crown-Rump Lengths

Volume 4: The Human Brain during the First Trimester 21- to 23-mm Crown-Rump Lengths

Volume 5: The Human Brain during the First Trimester 31- to 33-mm Crown-Rump Lengths

Volume 6: The Human Brain during the First Trimester 40- to 42-mm Crown-Rump Lengths

Volume 7: The Human Brain during the First Trimester 57- to 60-mm Crown-Rump Lengths

Volume 8: The Human Brain during the Second Trimester 96- to 150-mm Crown-Rump Lengths

Volume 9: The Human Brain during the Second Trimester 160- to 170-mm Crown-Rump Lengths

Volume 10: The Human Brain during the Second Trimester 190- to 210-mm Crown-Rump Lengths

Volume 11: The Human Brain during the Third Trimester 225- to 235-mm Crown-Rump Lengths

Volume 12: The Human Brain during the Third Trimester 260- to 270-mm Crown-Rump Lengths

Volume 13: The Human Brain during the Third Trimester 310- to 350-mm Crown-Rump Lengths

Volume 14: The Spinal Cord during the First Trimester

Volume 15: The Spinal Cord during the Second and Third Trimesters and the Early Postnatal Period

The Human Brain during the Second Trimester 190- to 210-mm Crown-Rump Lengths

Atlas of Human Central Nervous System Development, Volume 10

Shirley A. Bayer
Joseph Altman

CRC Press
Taylor & Francis Group
Boca Raton London New York

CRC Press is an imprint of the
Taylor & Francis Group, an **informa** business

First edition published 2024
by CRC Press
6000 Broken Sound Parkway NW, Suite 300, Boca Raton, FL 33487-2742

and by CRC Press
4 Park Square, Milton Park, Abingdon, Oxon, OX14 4RN

CRC Press is an imprint of Taylor & Francis Group, LLC

LCCN no. 2022008216

ISBN: 978-1-032-21940-0 (hbk)
ISBN: 978-1-032-21939-4 (pbk)
ISBN: 978-1-003-27067-6 (ebk)

DOI: 10.1201/9781003270676

Typeset in Times Roman
by KnowledgeWorks Global Ltd.

Access the Support Material: www.routledge.com/9781032219400

CONTENTS

ACKNOWLEDGMENTS

We thank the late Dr. William DeMyer, pediatric neurologist at Indiana University Medical Center, for access to his personal library on human CNS development. We also thank the staff of the National Museum of Health and Medicine that were at the Armed Forces Institute of Pathology, Walter Reed Hospital, Washington, D.C. when we collected data in 1995 and 1996: Dr. Adrianne Noe, Director; Archibald J. Fobbs, Curator of the Yakovlev Collection; Elizabeth C. Lockett; and William Discher. We are most grateful to the late Dr. James M. Petras at the Walter Reed Institute of Research who made his darkroom facilities available so that we could develop all the photomicrographs on location rather than in our laboratory in Indiana. Finally, we thank Chuck Crumly, Neha Bhatt, Kara Roberts, Michele Dimont, and Rebecca Condit for expert help during production of the manuscript.

AUTHORS

Shirley A. Bayer received her PhD from Purdue University in 1974 and spent most of her scientific career working with Joseph Altman. She was a professor of biology at Indiana-Purdue University in Indianapolis for several years, where she taught courses in human anatomy and developmental neurobiology while continuing to do research in brain development. Her lengthy publication record of dozens of peer-reviewed, scientific journal articles extends back to the mid 1970s. She has co-authored several books and many articles with her late spouse, Joseph Altman. It was her research (published in *Science* in 1982) that proved that new neurons are added to granule cells in the dentate gyrus during adult life, a unique neuronal population that grows. That paper stimulated interest in the dormant field of adult neurogenesis.

Joseph Altman, now deceased, was born in Hungary and migrated with his family via Germany and Australia to the US. In New York, he became a graduate student in psychology in the laboratory of Hans-Lukas Teuber, earning a PhD in 1959 from New York University. He was a postdoctoral fellow at Columbia University, and later joined the faculty at the Massachusetts Institute of Technology. In 1968, he accepted a position as a professor of biology at Purdue University. During his career, he collaborated closely with Shirley A. Bayer. From the early 1960s-2016, he published many articles in peer-reviewed journals, books, monographs, and free online books that emphasized developmental processes in brain anatomy and function. His most important discovery was adult neurogenesis, the creation of new neurons in the adult brain. This discovery was made in the early 1960s while he was based at MIT, but was largely ignored in favor of the prevailing dogma that neurogenesis is limited to prenatal development. After Dr. Bayer's paper proved new neurons are added to granule cells in the hippocampus, Dr. Altman's monumental discovery became more accepted. During the 1990s, new researchers "rediscovered" and confirmed his original finding. Adult neurogenesis has recently been proven to occur in the dentate gyrus, olfactory bulb, and striatum through the measurement of Carbon-14—the levels of which changed during nuclear bomb testing throughout the 20th century—in postmortem human brains. Today, many laboratories around the world are continuing to study the importance of adult neurogenesis in brain function. In 2011, Dr. Altman was awarded the Prince of Asturias Award, an annual prize given in Spain by the Prince of Asturias Foundation to individuals, entities, or organizations globally who make notable achievements in the sciences, humanities, and public affairs. In 2012, he received the International Prize for Biology - an annual award from the Japan Society for the Promotion of Science (JSPS) for "outstanding contribution to the advancement of research in fundamental biology." This Prize is one of the most prestigious honors a scientist can receive. When Dr. Altman died in 2016, Dr. Bayer continued the work they started over 50 years ago. In her late husband's honor, she created the Altman Prize, awarded each year by JSPS to an outstanding young researcher in developmental neuroscience.

INTRODUCTION

A. Specimens and Organization

Volume 10 in the *Atlas of Human Central Nervous System Development* series illustrates development of the 6-month-old human brain in the late second trimester (originally presented in Bayer and Altman 2005). Differentiation continues throughout the diencephalon, midbrain, pons, and medulla. These structures are growing and appear mature at low magnification. The telencephalon (cerebral cortex, basal ganglia) continue generating neurons, and complex migration patterns appear, especially in the ***stratified transitional field*** of the cerebral cortex. Immature features continue to predominate in the cerebellum where neurons are still migrating and its secondary germinal matrix, the ***external germinal layer***, continues to generate basket, stellate, and granule cells.

Grayscale photographs of Nissl-stained sections of three normal brains in the Yakovlev Collection[1] are illustrated. **Part II** presents Y197-65 (190-mm crown-rump length) cut in the horizontal plane. Horizontal plates are ordered from superior to inferior; the anterior part of each section is facing to the left, posterior to the right, and the midline is in the horizontal center of each section. **Part III** presents Y24-60 (205-mm crown-rump length) cut in the sagittal plane. Plates are presented in serial order from medial to lateral; the superior (dorsal) part of each section is at the top of the page, the inferior (ventral) part at the bottom. **Part IV** presents Y94-62 (210-mm crown-rump length) cut in the frontal plane. Plates are presented in serial order from anterior to posterior; dorsal (superior) part of the section is on the top, inferior part on the bottom, and the midline is in the vertical center. **Part A** of each plate on the left page shows the full contrast photograph without labels; **Part B** shows low contrast copies of the same photograph on the right page with superimposed outlines of structures and unabbreviated labels. The *low-magnification plates* show entire sections to identify large structures of the brain. The *high-magnification plates* feature enlarged views of the brain core to identify smaller structures. A few *very-high-magnification plates* show the cerebral cortex and cerebellum in great detail. As in the other volumes, transient structures are labeled in ***italics***. During fixation, shrinkage introduced artifactual infolding of the cerebral cortex in some specimens. During dissection, embedding, cutting, and staining, some of the sections illustrated were torn. Both artifacts and processing damage are usually outlined with *dashed lines* in Part **B** of each plate.

1. The *Yakovlev Collection* (designated by a **Y** prefix in the specimen number) is the work of Dr. Paul Ivan Yakovlev (1894–1983), a neurologist affiliated with Harvard University and the AFIP. Throughout his career, Yakovlev collected many diseased and normal human brains. He invented a giant microtome that was capable of sectioning entire human brains. Later, he became interested in the developing brain and collected many human brains during the second and third trimesters. The normal brains in the developmental group were cataloged by Haleem (1990) and were examined by us during 1996 and 1997. The collection was moved to the National Museum of Health and Medicine when the Armed Forces Institute of Pathology (AFIP) closed and is still available for research.

B. Developmental Highlights

Figures 1-3 focus on the ***early lobulation*** in the cerebral cortex because it begins to enhance during the second trimester and will continue throughout the rest of brain maturation into adulthood. Illustrations are modified from our book on the *Development of the Human Neocortex* (Altman and Bayer, 2015), that is available free online at neurondevelopment.org.

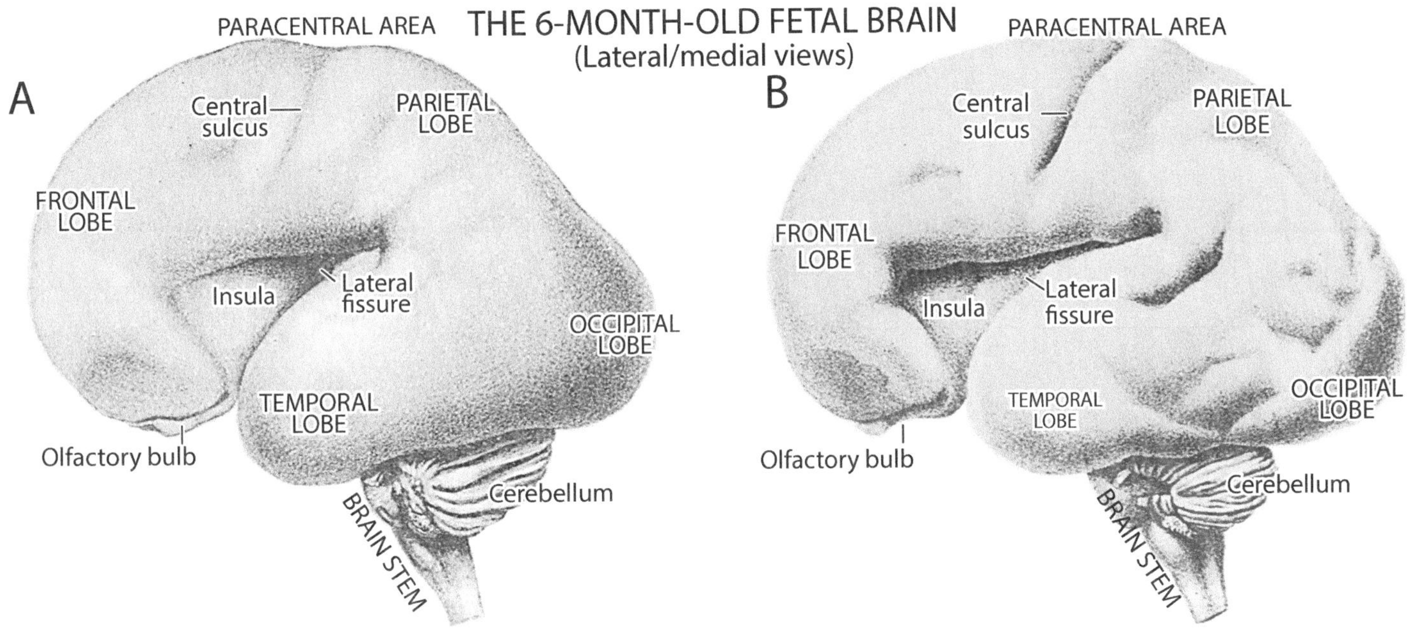

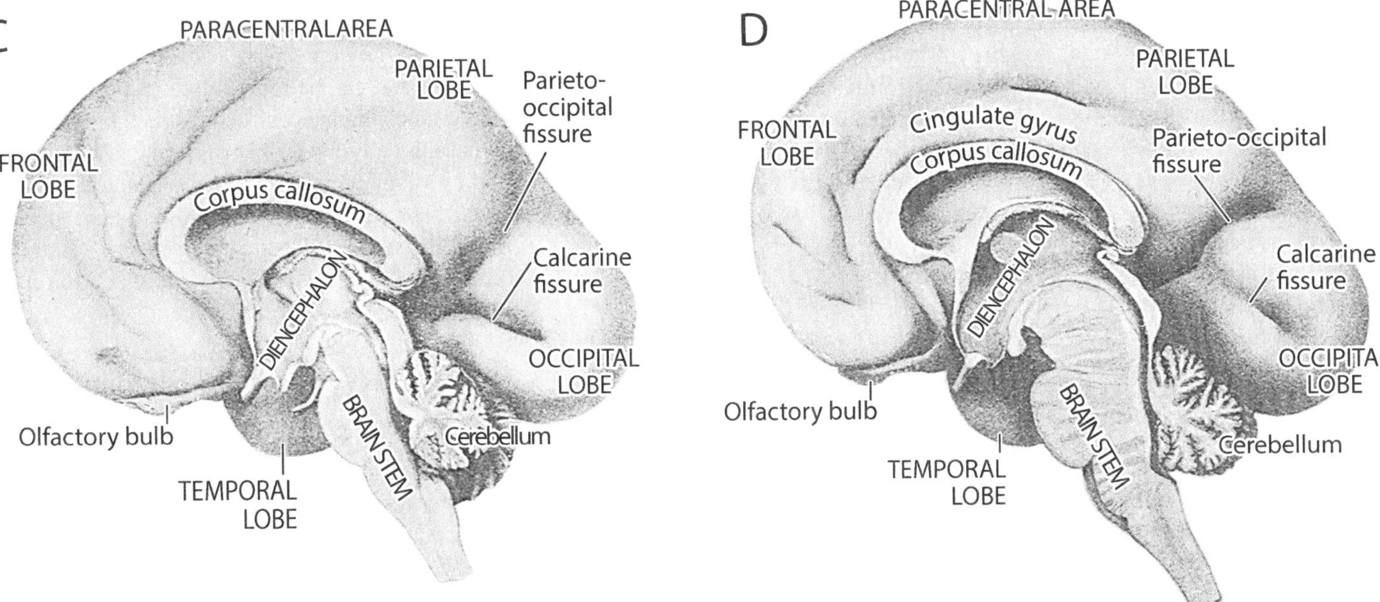

Figure 1 (facing pages). The brains of early 6-month-old fetuses (left columns, A, C, E, G) and late 6-month-old fetuses (right columns, B, D, F, H) in lateral (top row, A, B) medial (bottom row, C, D), superior (top row, E, F), and inferior (G, H) views. (After Retzius, 1896) The focus of this illustration is to point out the early sulci and fissures of the fetal brain that start to elaborate in the second trimester. Major fissures appear before that. The **interhemispheric fissure** is the deep midline cleft, first to appear at the 10.5-mm crown-rump length (Bayer and Altman, 2023a, Volume 2, Part V). The **lateral fissure** is next,

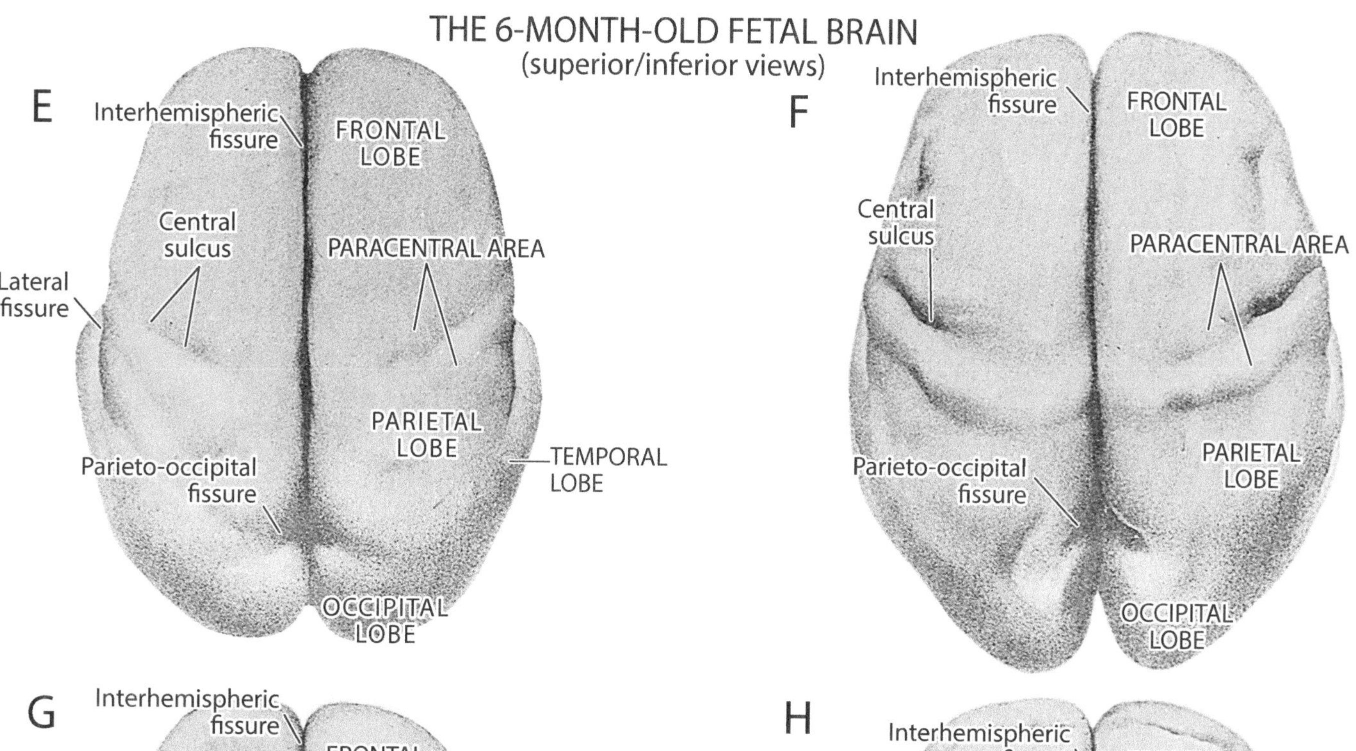

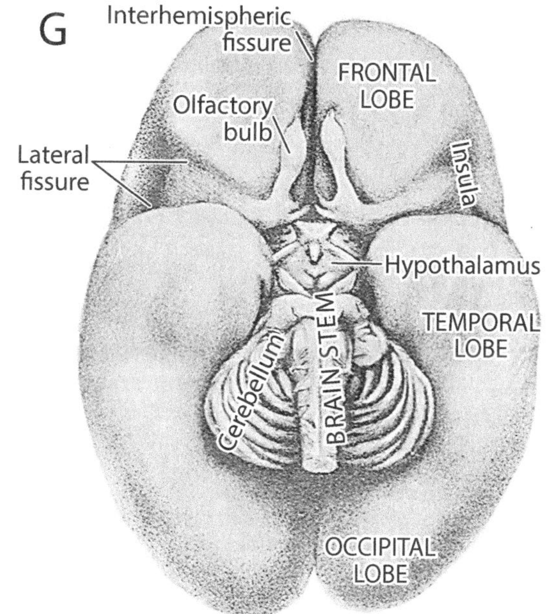

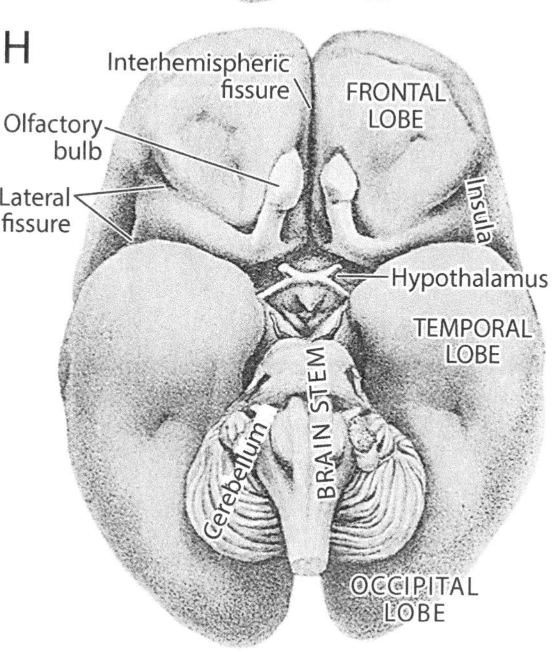

first appearing as a lateral indentation of the insular cortex between the frontal and temporal areas at the 32-mm crown-rump length (Bayer and Altman,2023c, Volume 5, Part II). All other early fissures and sulci appear during the second trimester. The **calcarine fissure** appears at the 96-mm crown-rump length in the 3.5-month-old fetal brain. It is a prominent indentation in the midline of the posterior neocortex, the presumptive visual area in the occipital lobe (Bayer and Altman, in press, Volume 8, Parts II-IV). The **cingulate** and **central** sulci, and the **parieto-occipital fissure** appear at 160 to 165-mm crown-rump length (Bayer and Altman, in press b, Volume 9, Parts II-III).
(Modified Figure 21 in Altman and Bayer, 2015.)

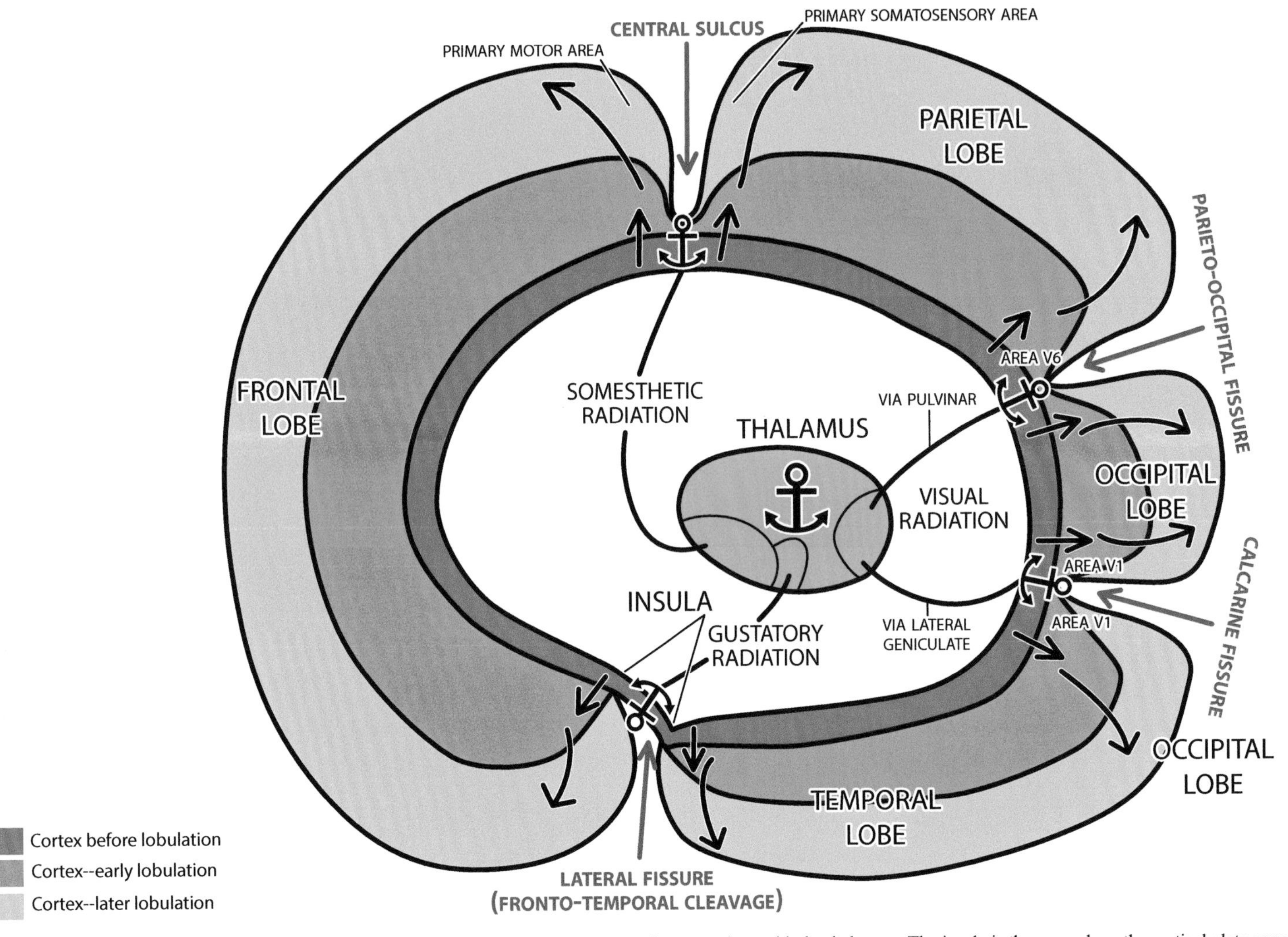

Figure 2. This diagram shows the basic elements of a hypothesis that cortical lobulation is initiated by early connections with the thalamus. The insula is the area where the cortical plate appears first at the 23-mm crown-rump length (Bayer and Altman, 2023b, Volume 4, Part III). It receives imput from the gustatory area. The insular cortex is also important in emotional behavior, and eventually will be covered by the posterior growth of the frontal lobe and the forward growth of the temporal lobe. The depths of the calcarine and parieto-occipital fissures are established by the dorsal visual radiation (via the pulvinar) and the ventral radiation (via the lateral geniculate). The thalamic somesthetic radiation establishes the depths of the central sulcus that first starts laterally in the second trimester and advances medially during the third trimester. Literature citing the early anatomical connections are discussed at length in Altman and Bayer (2015). (Figure 81 in Altman and Bayer, 2015.)

REFERENCES

Altman J, Bayer SA. (2015) *Development of the Human Neocortex.* Ocala, FL, Laboratory of Developmental Neurobiology, neurondevelopment.org.

Bayer SA, Altman J (2005) *Atlas of Human Central Nervous System Development*, Volume 3: *The Human Bran during the Second Trimester.* Boca Raton, FL, CRC Press.

Bayer SA, Altman J (2023a) *The Human Brain during the First Trimester 6.3- to 10.5-mm Crown-Rump Lengths, Atlas of Human Central Nervous System Development*, Volume 2. Taylor and Francis, CRC Press.

Bayer SA, Altman J (2023b) *The Human Brain during the First Trimester 21- to 23-mm Crown-Rump Lengths, Atlas of Human Central Nervous System Development*, Volume 4. Taylor and Francis, CRC Press.

Bayer SA, Altman J (2023c) *The Human Brain during the First Trimester 31- to 33-mm Crown-Rump Lengths, Atlas of Human Central Nervous System Development*, Volume 5. Taylor and Francis, CRC Press.

Bayer SA, Altman J (in press b) *The Human Brain during the Second Trimester 96- to 150-mm Crown-Rump Lengths, Atlas of Human Central Nervous System Development*, Volume 8. Taylor and Francis, CRC Press.

Bayer SA, Altman J (in press c) *The Human Brain during the Second Trimester 160- to 170-mm Crown-Rump Lengths, Atlas of Human Central Nervous System Development*, Volume 9. Taylor and Francis, CRC Press.

Gamberini MS, Bakola S, Passarelli L, et al. (2016) Thalamic projections to visual and visuomotor areas (V6 and V6A) in the rostral bank of the parieto-occipital sulcus of the macaque. *Brain Structure and Function*, 221:1573-1589.

Haleem M (1990) *Diagnostic Categories of the Yakovlev Collection of Normal and Pathological Anatomy and Development of the Brain.* Washington, D.C. Armed Forces Institute of Pathology.

Retzius, G. (1896) *Das Menschenhirn: Studien in der makroskopischen Morphologie.* Vols. 1 and 2. Stockholm: Königliche Buchdruckerei.

PART II: Y197-65
CR 170 mm (GW 23)
Horizontal

This specimen is a stillborn female fetus (Yakovlev case number RPSL B-197-65) with a crown rump length (CR) of 190 mm estimated to be at gestational week (GW) 23. The brain was cut in the horizontal plane in 939 sections, 35-μm thick and is classified as a Normative Control in the Yakovlev Collection (Haleem, 1990). Since there is no photograph of this brain before it was embedded and cut, we turned to the atlas that Retzius published in 1896 showing whole fetal brains in medial, lateral, superior, and inferior views and midline sagittally cut brains. **Figure 3**, from Retzius (1896), shows a lateral view of a GW 24 brain that is similar to the features of Y197-65, along with the approximate cutting angle of the sections. Photographs of 16 Nissl-stained sections are illustrated. **Plates 1–8** show low-magnification views of the large sections that contain the entire cerebral cortex. **Plates 9–16** are high-magnification views of the brain core with parts of the surrounding cortex. **Plates 17–23** show high-magnification views of the brainstem only.

In the cortical regions of the telencephalon, the ***neuroepithelium/subventricular zone*** is generating neocortical neurons destined to reside in superficial cortical layers (mainly II and III). Many migrating and sojourning neurons and/or glia, with interdigitated fibers, are visible in all lobes of the cerebral cortex as ***stratified transitional fields***, with pronounced regional heterogeneity between granular (future sensory) and agranular (future motor) areas. In lateral parts of the cerebral cortex, streams of neurons and/or glia are numerous in the ***lateral migratory stream*** heading toward the insular cortex, primary olfactory cortex, temporal cortex, and basolateral parts of the amygdaloid complex. In the hippocampus, cells are entering Ammon's horn pyramidal layer via the ***ammonic migratory stream***, and granule cells and their precursors are migrating to the dentate gyrus in the ***dentate migratory stream***. Stem cells form a ***subgranular zone*** in the hilus and generate granule cells that migrate into the overlying dentate granular layer. There is a prominent ***neuroepithelium/subventricular zone*** in the nucleus accumbens and striatum where neurons (and glia) are being generated. The striatal portion can be subdivided into anterolateral, anteromedial, and posterior parts. The septum has a ***glioepithelium/ependyma*** at the ventricle. Many neurons, glia, and their mitotic precursor cells are still migrating through the olfactory peduncle toward the olfactory bulb (***rostral migratory stream***) from a presumed source area in the germinal matrix at the junction between the cerebral cortex and striatum (***cortical/striatal neuroepithelium and subventricular zone***).

Most of the structures in the diencephalon appear to be settled and are maturing, and the third ventricle is lined by a thin ***glioepithelium/ependyma***. However, lamination in the lateral geniculate body is just beginning. In the midbrain, pons, and medulla, there is a convoluted ***glioepithelium/ependyma*** lining the cerebral aqueduct and fourth ventricle. There is an extensive ***raphe migration*** in the midline of the posterior pons/anterior medulla blending with the arcuate nucleus on the ventromedial surface of the pyramids.

The dentate nucleus in the cerebellum is partially laminated. The entire surface of the cerebellar cortex is covered by the prominent ***external germinal layer*** that is actively producing basket, stellate, and granule cells. Most cortical neurons, including Purkinje cells, are still migrating. Lobulation is continuing in both the vermis and hemispheres. The ***germinal trigone*** is large at the base of the nodulus and along the floccular peduncle.

GW 23 HORIZONTAL SECTION PLANES

Figure 3. Lateral view of a GW 23 brain with the approximate locations and cutting angles of the sections of Y197-65 (Figure 4 in Table 11, Volume 2, Retzius, 1896).

PLATE 1A
CR 190 mm
GW 23, Y197-65
Horizontal
Section 261

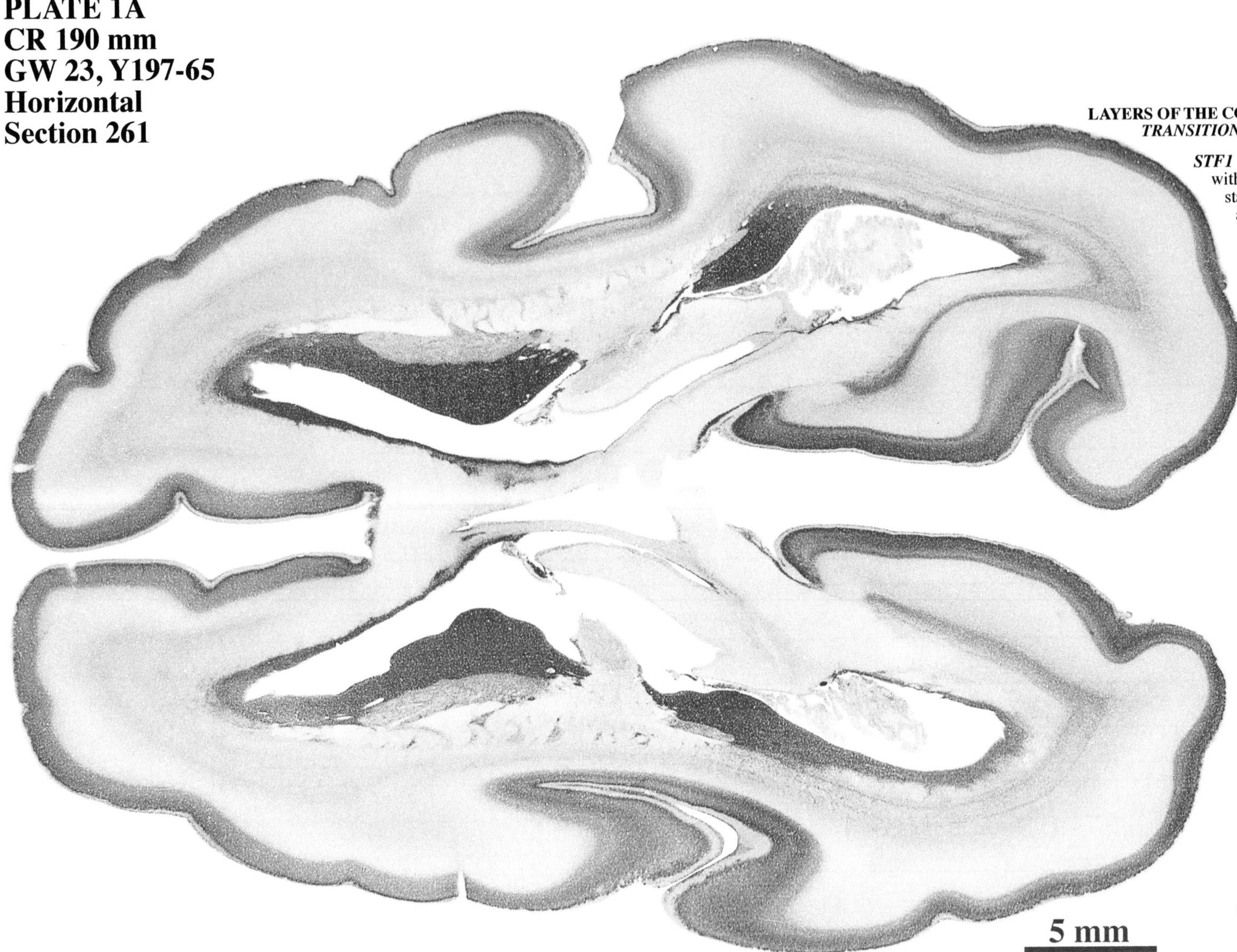

LAYERS OF THE CORTICAL *STRATIFIED TRANSITIONAL FIELD (STF)*

STF1 Superficial fibrous layer with an early developmental stage ***(t1)*** when many cells are migrating through it, followed by a late stage ***(t2)*** with sparse cells. Endures as the subcortical white matter.

STF2 Upper cellular layer, the last sojourn zone before cells translocate to the cortical plate.

STF3 Honeycomb trilaminar matrix ***(3a, 3b, 3c)*** of cells and fibers found only in granular cortices.

STF4 Complex middle layer with three developmental stages:
t1– fibrous layer without interspersed cells; ***t2***– cells and fibers intermingle to form striations;
t3– fibers endure in the deep white matter.

STF5 Deep cellular layer, the first sojourn zone to appear outside the germinal matrix.

STF6 Late-forming deep layer of callosal fibers outside the germinal matrix.

PLATE 1B

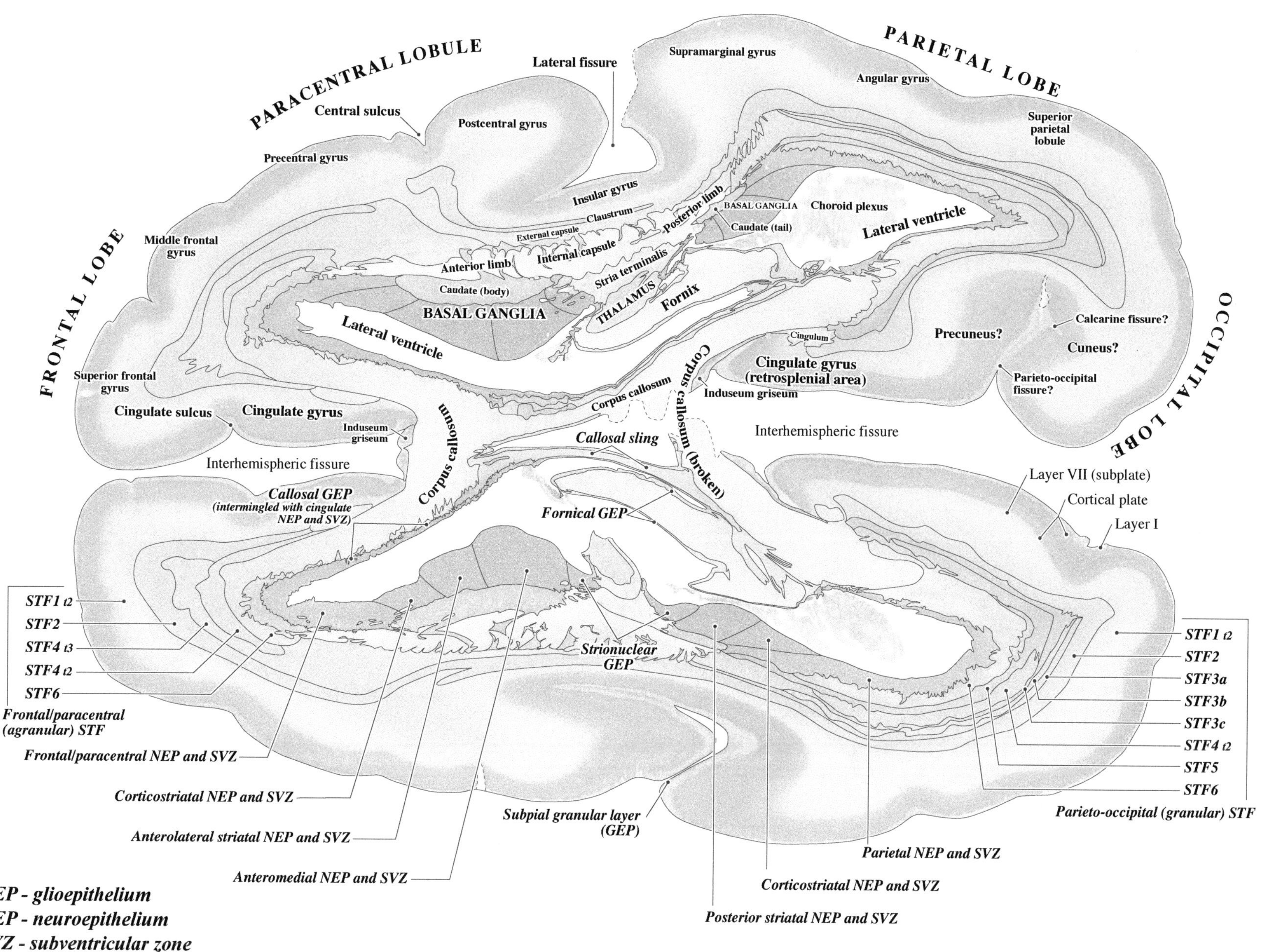

GEP - glioepithelium
NEP - neuroepithelium
SVZ - subventricular zone

Germinal and transitional structures in *italics*

PLATE 2A
CR 190 mm
GW 23, Y197-65
Horizontal
Section 301

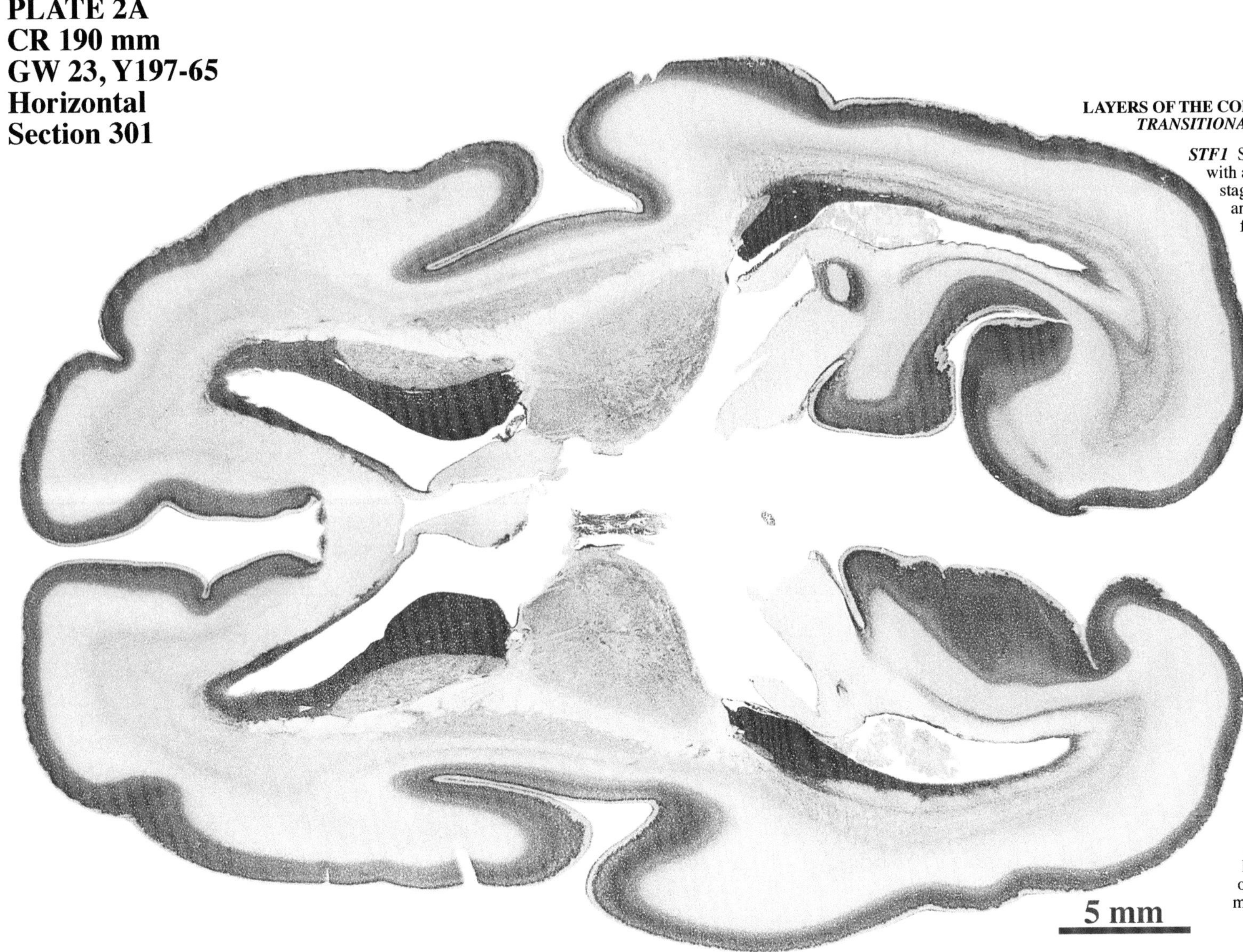

LAYERS OF THE CORTICAL *STRATIFIED TRANSITIONAL FIELD (STF)*

STF1 Superficial fibrous layer with an early developmental stage ***(t1)*** when many cells are migrating through it, followed by a late stage ***(t2)*** with sparse cells. Endures as the subcortical white matter.

STF2 Upper cellular layer, the last sojourn zone before cells translocate to the cortical plate.

STF3 Honeycomb trilaminar matrix ***(3a, 3b, 3c)*** of cells and fibers found only in granular cortices.

STF4 Complex middle layer with three developmental stages:
t1– fibrous layer without interspersed cells; ***t2***– cells and fibers intermingle to form striations;
t3– fibers endure in the deep white matter.

STF5 Deep cellular layer, the first sojourn zone to appear outside the germinal matrix.

STF6 Late-forming deep layer of callosal fibers outside the germinal matrix.

See detail of the brain core in Plates 9A and B.

PLATE 2B

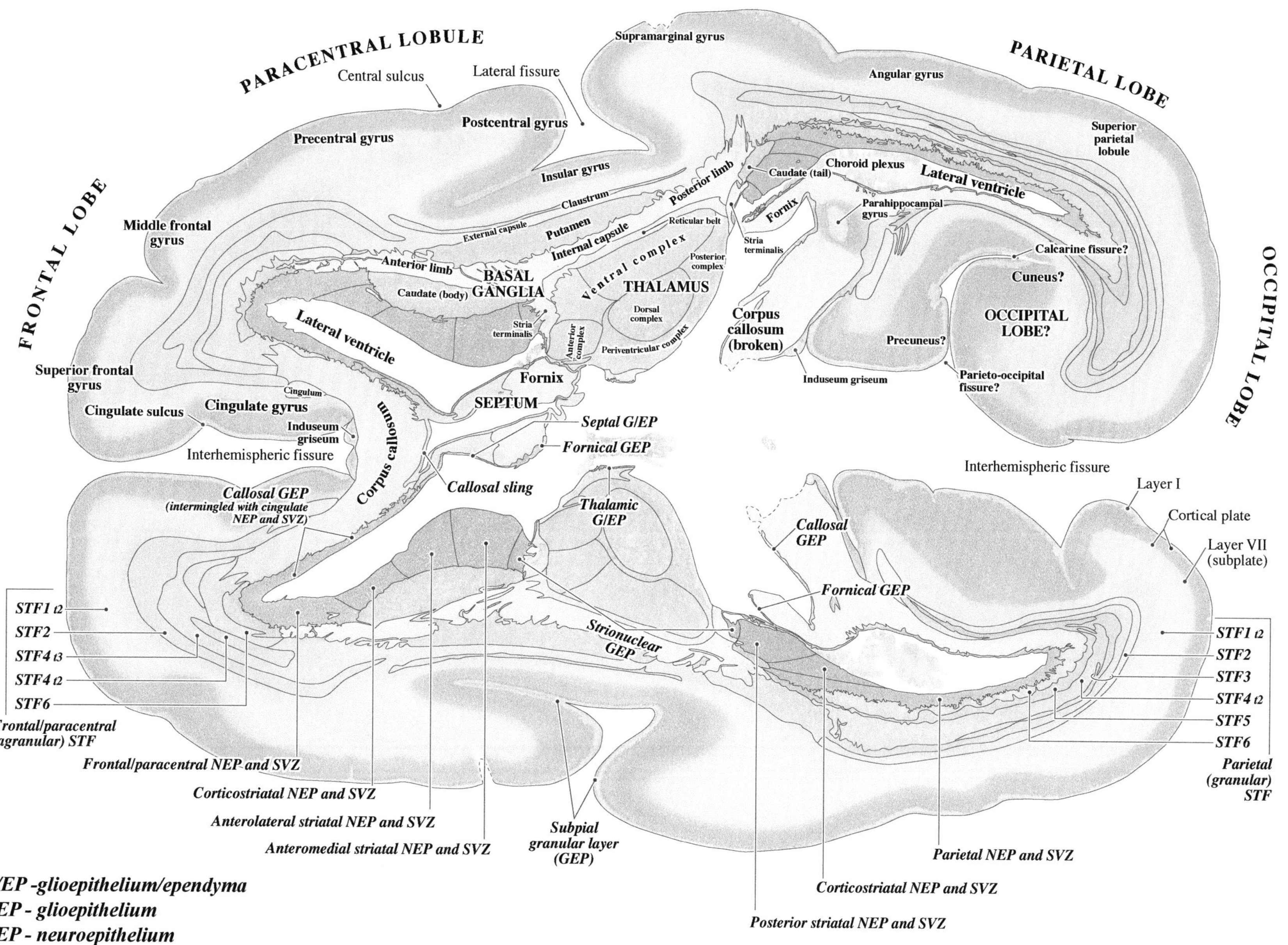

G/EP -glioepithelium/ependyma
GEP - glioepithelium
NEP - neuroepithelium
SVZ - subventricular zone
Germinal and transitional structures in *italics*

PLATE 3A
CR 190 mm
GW 23, Y197-65
Horizontal
Section 361

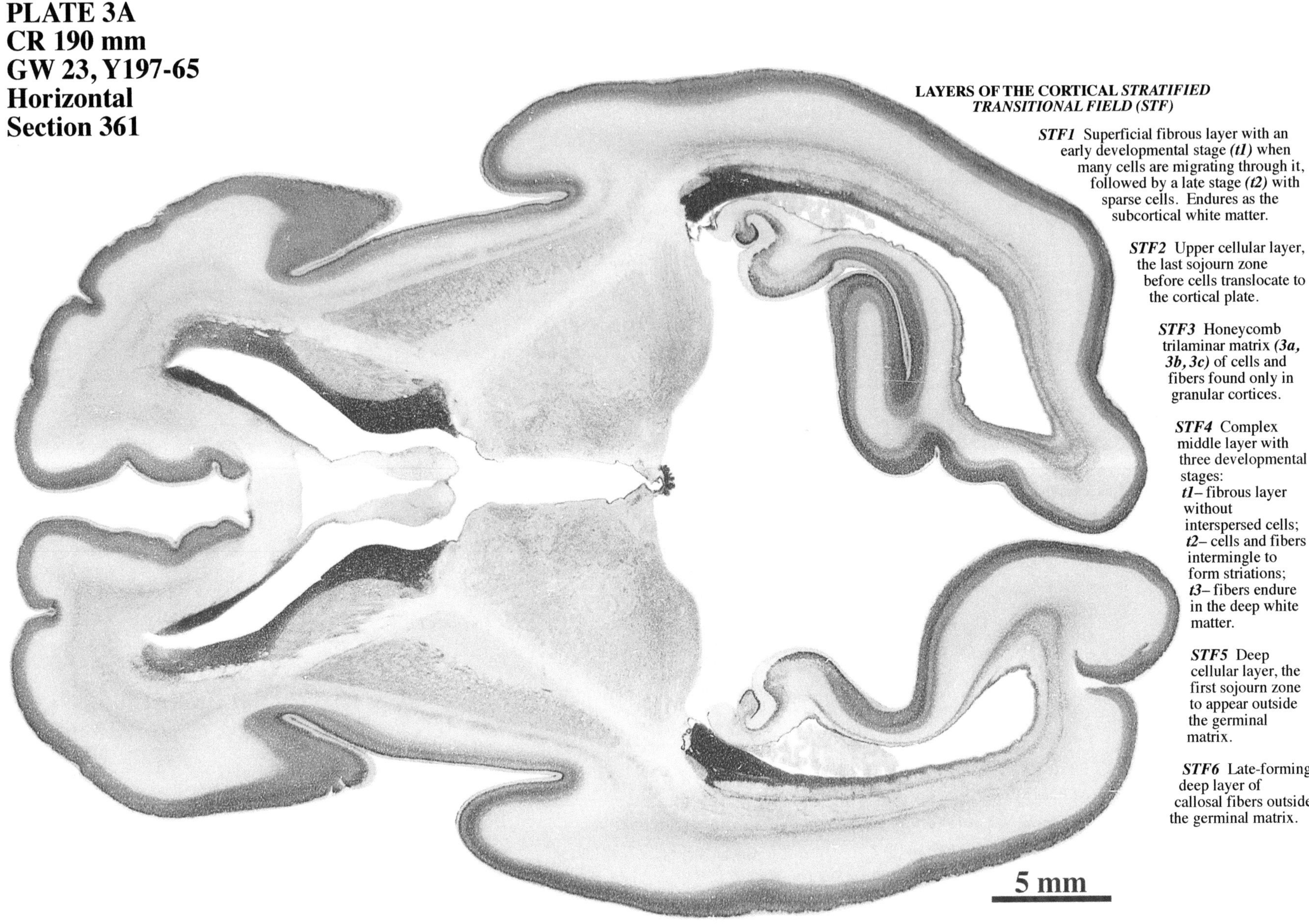

See detail of the brain core in Plates 10A and B.

PLATE 3B

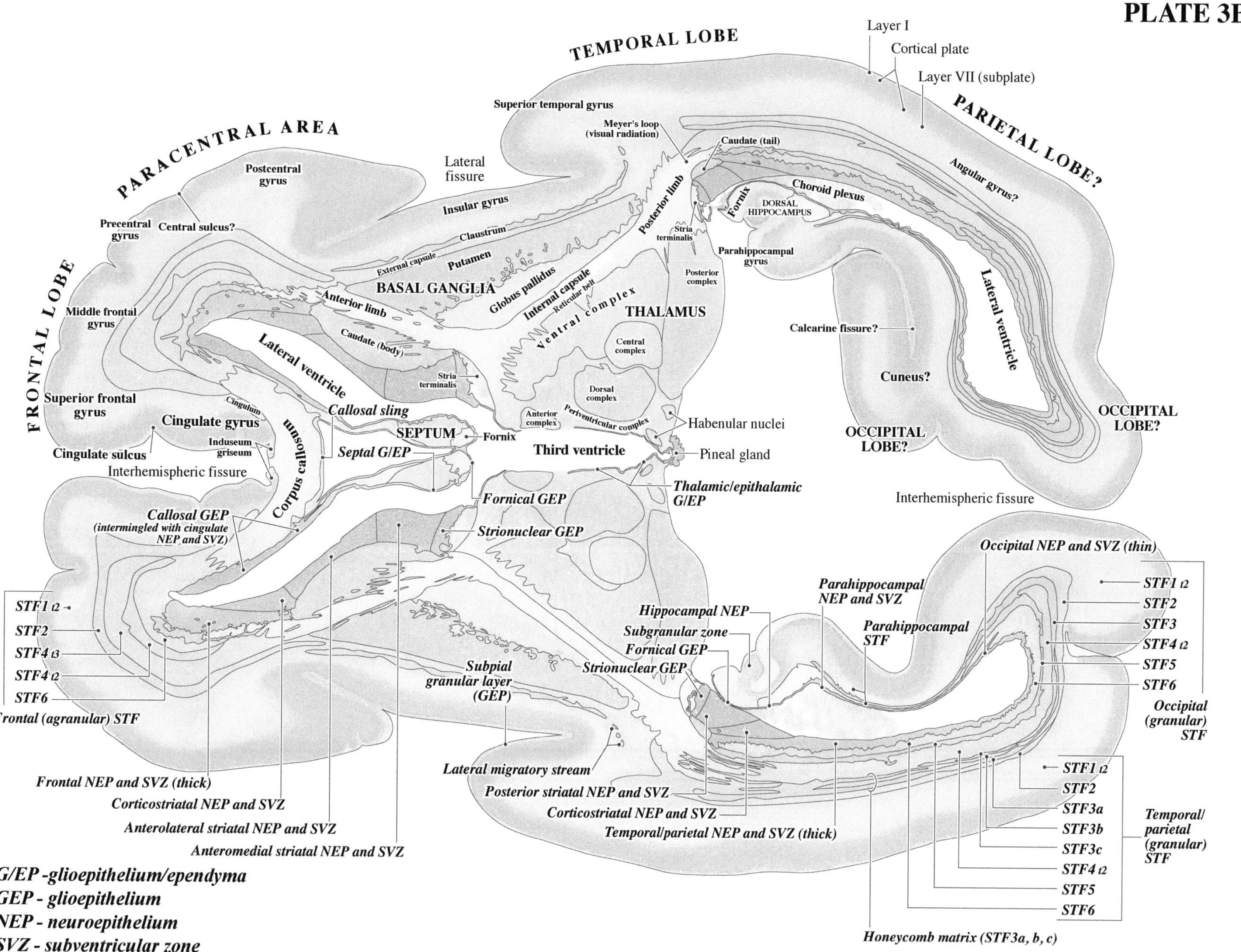

G/EP* -*glioepithelium/ependyma

GEP* - *glioepithelium

NEP* - *neuroepithelium

SVZ* - *subventricular zone

Germinal and transitional structures in *italics*

PLATE 4A
CR 190 mm
GW 23, Y197-65
Horizontal
Section 431

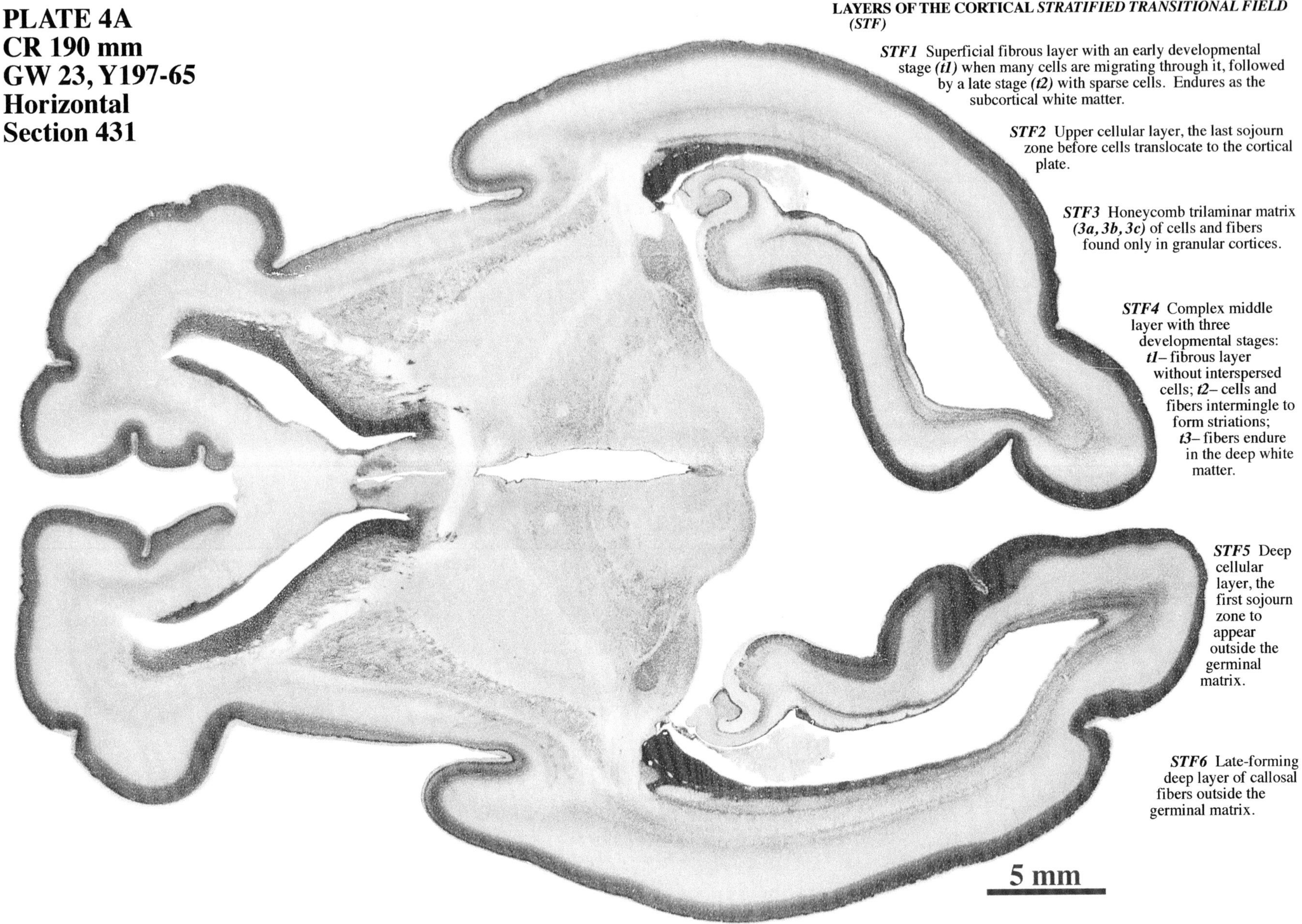

LAYERS OF THE CORTICAL *STRATIFIED TRANSITIONAL FIELD (STF)*

STF1 Superficial fibrous layer with an early developmental stage ***(t1)*** when many cells are migrating through it, followed by a late stage ***(t2)*** with sparse cells. Endures as the subcortical white matter.

STF2 Upper cellular layer, the last sojourn zone before cells translocate to the cortical plate.

STF3 Honeycomb trilaminar matrix ***(3a, 3b, 3c)*** of cells and fibers found only in granular cortices.

STF4 Complex middle layer with three developmental stages: ***t1***– fibrous layer without interspersed cells; ***t2***– cells and fibers intermingle to form striations; ***t3***– fibers endure in the deep white matter.

STF5 Deep cellular layer, the first sojourn zone to appear outside the germinal matrix.

STF6 Late-forming deep layer of callosal fibers outside the germinal matrix.

See detail of the brain core in Plates 11A and B.

PLATE 4B

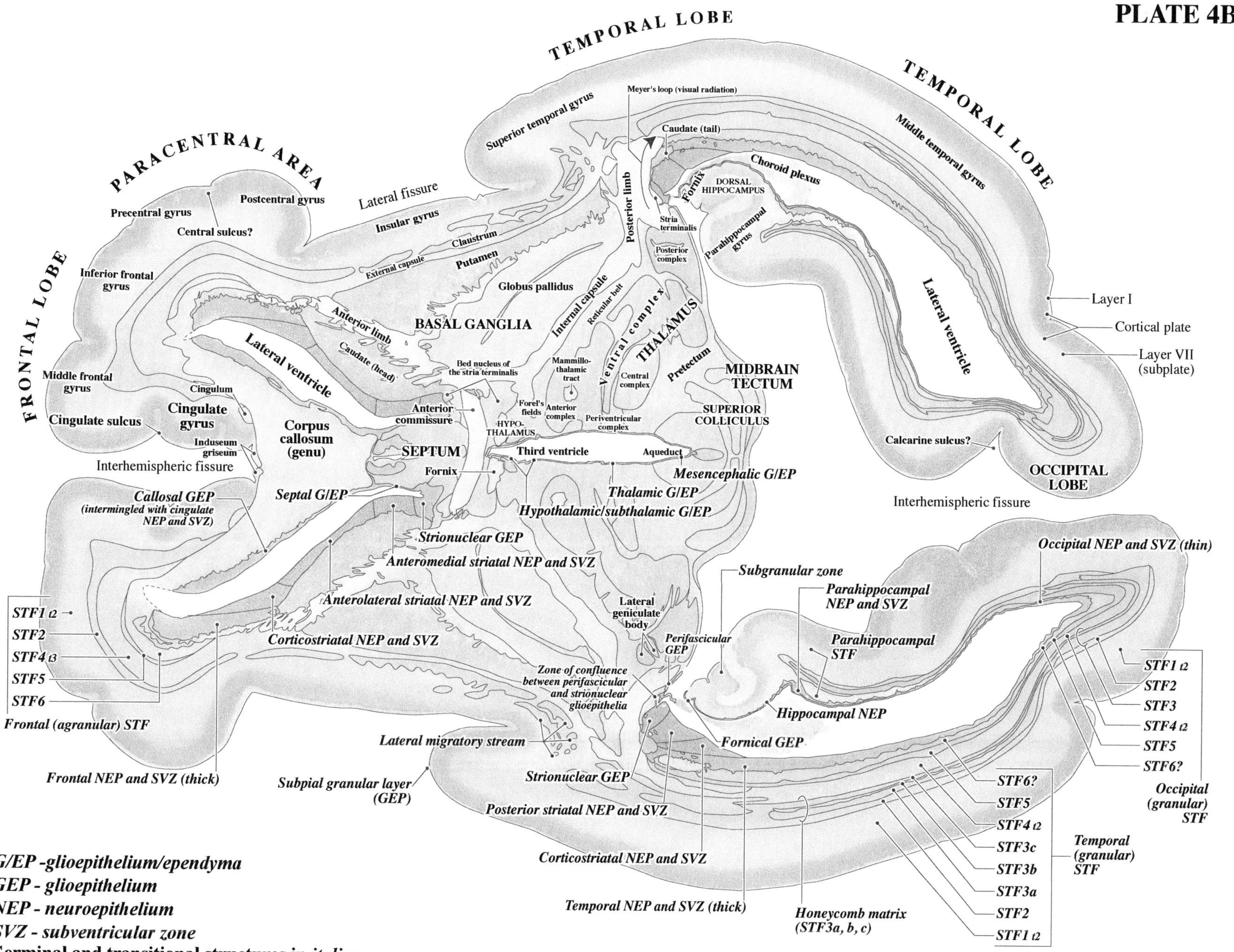
TEMPORAL LOBE
TEMPORAL LOBE
PARACENTRAL AREA
FRONTAL LOBE
Superior temporal gyrus
Meyer's loop (visual radiation)
Caudate (tail)
Middle temporal gyrus
Choroid plexus
Fornix
DORSAL HIPPOCAMPUS
Posterior limb
Stria terminalis
Posterior complex
Parahippocampal gyrus
Lateral fissure
Precentral gyrus
Postcentral gyrus
Central sulcus?
Insular gyrus
Claustrum
External capsule
Putamen
Inferior frontal gyrus
Globus pallidus
Internal capsule
Reticular belt
Ventral complex
THALAMUS
Lateral ventricle
BASAL GANGLIA
Anterior limb
Caudate (head)
Lateral ventricle
Layer I
Cortical plate
Layer VII (subplate)
Middle frontal gyrus
Cingulum
Cingulate gyrus
Cingulate sulcus
Induseum griseum
Corpus callosum (genu)
Bed nucleus of the stria terminalis
Anterior commissure
Mammillo-thalamic tract
Forel's fields
Anterior complex
HYPO-THALAMUS
Periventricular complex
Central complex
Pretectum
MIDBRAIN TECTUM
SUPERIOR COLLICULUS
Calcarine sulcus?
OCCIPITAL LOBE
Interhemispheric fissure
SEPTUM
Third ventricle
Aqueduct
Fornix
Mesencephalic G/EP
Thalamic G/EP
Hypothalamic/subthalamic G/EP
Interhemispheric fissure
Callosal GEP (intermingled with cingulate NEP and SVZ)
Septal G/EP
Strionuclear GEP
Anteromedial striatal NEP and SVZ
Occipital NEP and SVZ (thin)
Subgranular zone
Parahippocampal NEP and SVZ
Anterolateral striatal NEP and SVZ
Lateral geniculate body
STF1 t2
STF2
STF4 t3
STF5
STF6
Frontal (agranular) STF
Corticostriatal NEP and SVZ
Perifascicular GEP
Parahippocampal STF
Zone of confluence between perifascicular and strionuclear glioepithelia
Hippocampal NEP
STF1 t2
STF2
STF3
STF4 t2
STF5
STF6?
Lateral migratory stream
Fornical GEP
Frontal NEP and SVZ (thick)
Subpial granular layer (GEP)
Strionuclear GEP
Posterior striatal NEP and SVZ
STF6?
STF5
STF4 t2
STF3c
STF3b
STF3a
STF2
STF1 t2
Occipital (granular) STF
Temporal (granular) STF
Corticostriatal NEP and SVZ
Temporal NEP and SVZ (thick)
Honeycomb matrix (STF3a, b, c)
G/EP -glioepithelium/ependyma
GEP - glioepithelium
NEP - neuroepithelium
SVZ - subventricular zone
Germinal and transitional structures in italics

PLATE 5A
CR 190 mm
GW 23, Y197-65
Horizontal
Section 491

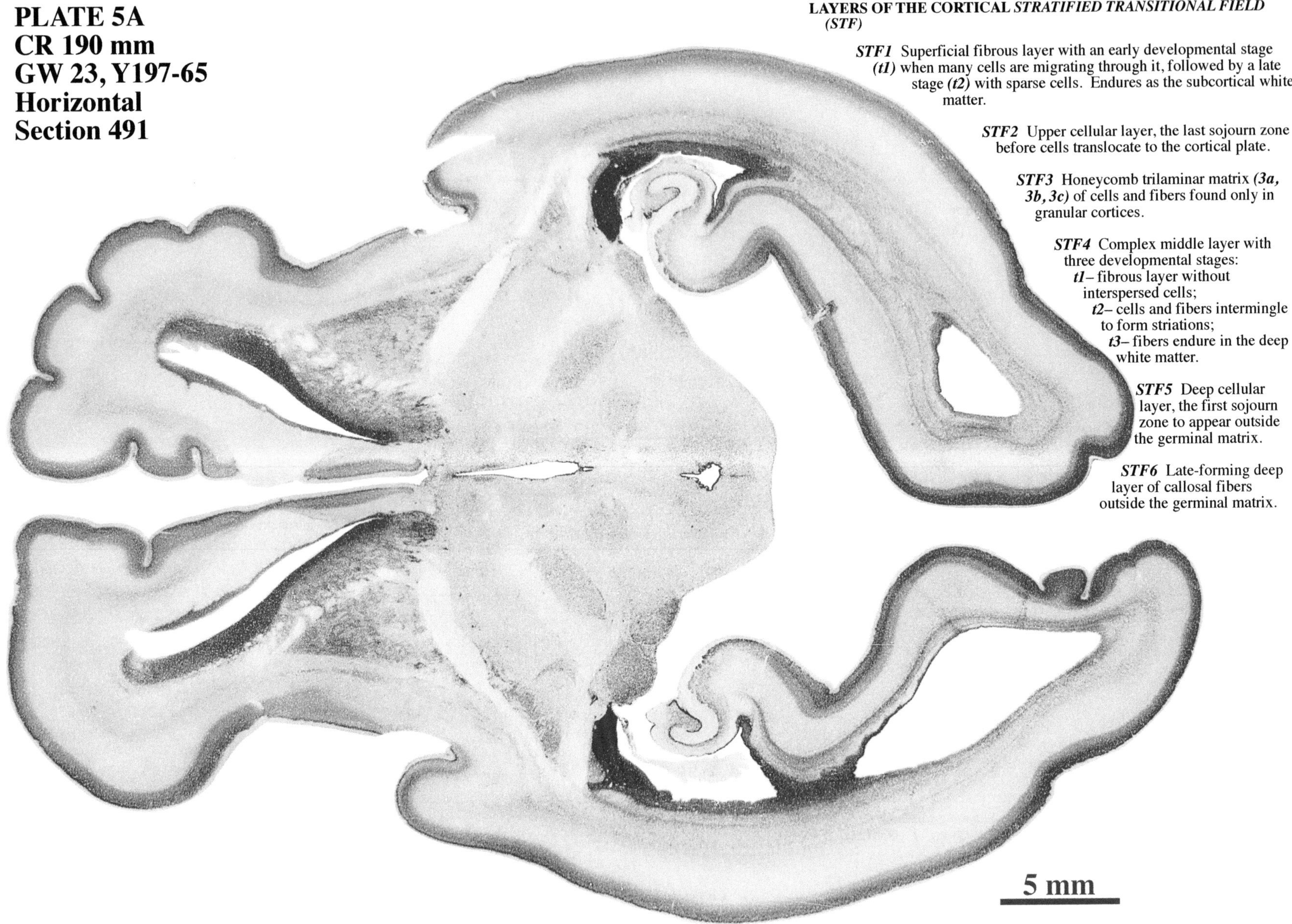

See detail of the brain core in Plates 12A and B.

PLATE 5B

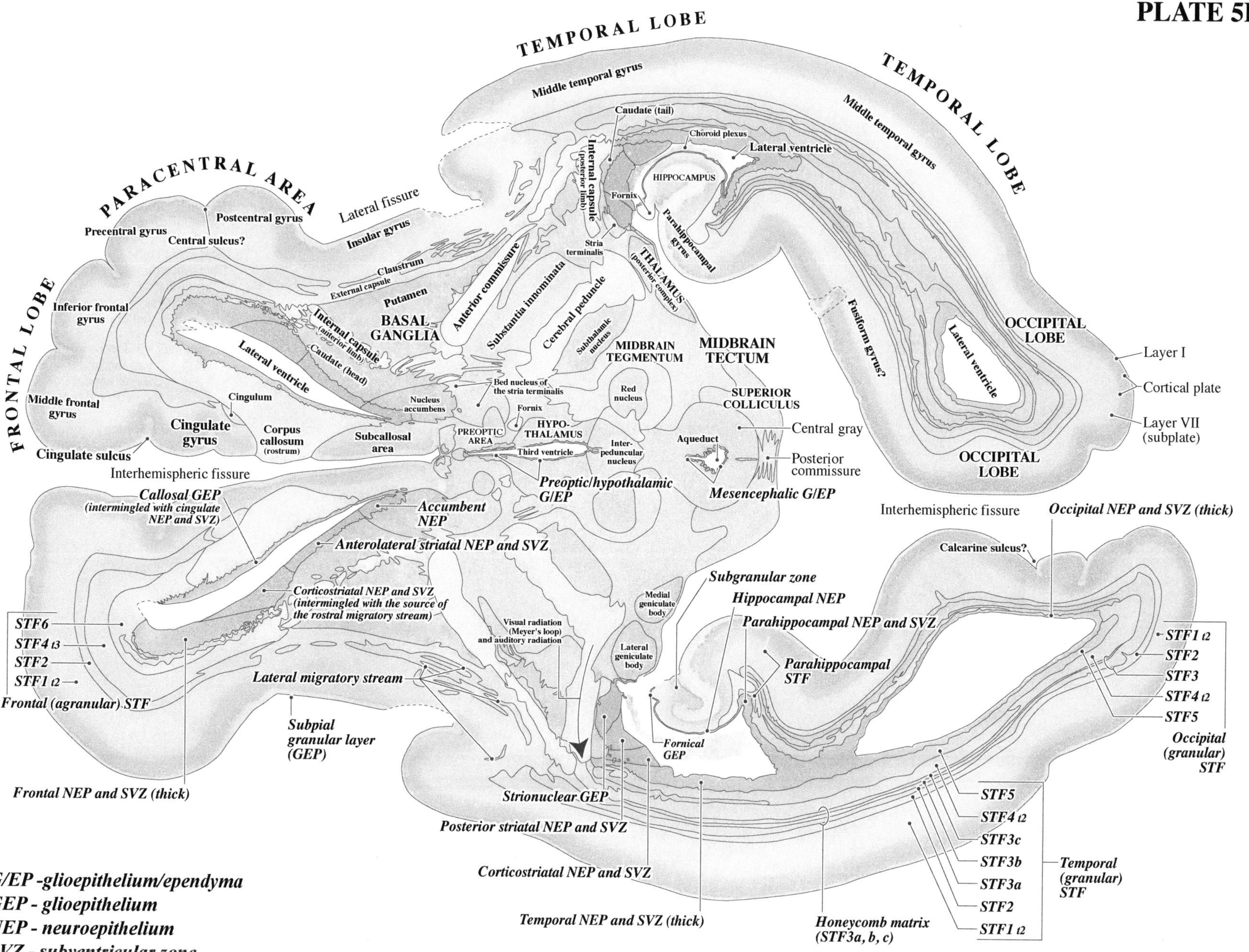

G/EP -glioepithelium/ependyma
GEP - glioepithelium
NEP - neuroepithelium
SVZ - subventricular zone
Germinal and transitional structures in *italics*

PLATE 6A
CR 190 mm
GW 23, Y197-65
Horizontal
Section 521

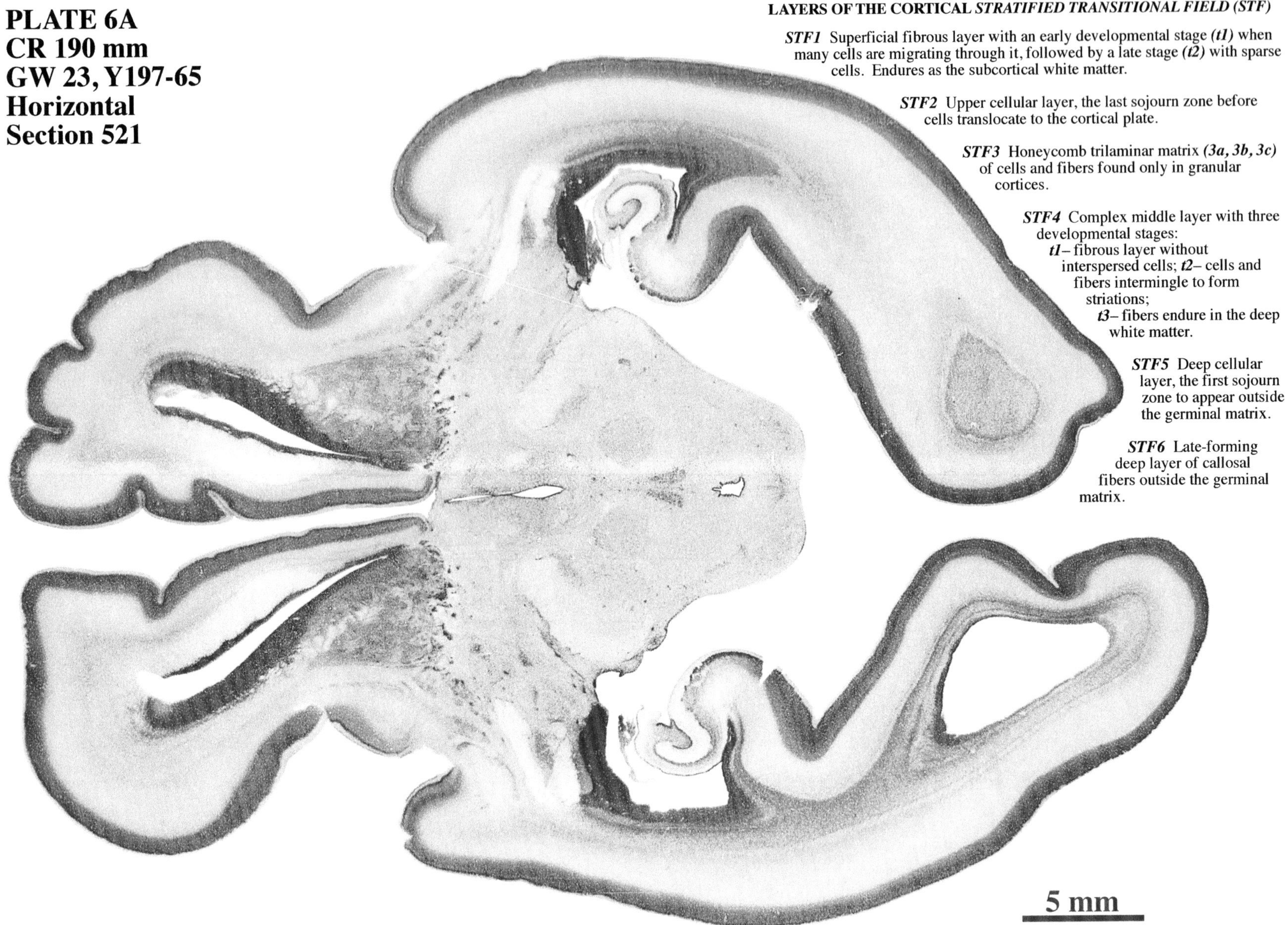

See detail of the brain core in Plates 13A and B.

PLATE 6B

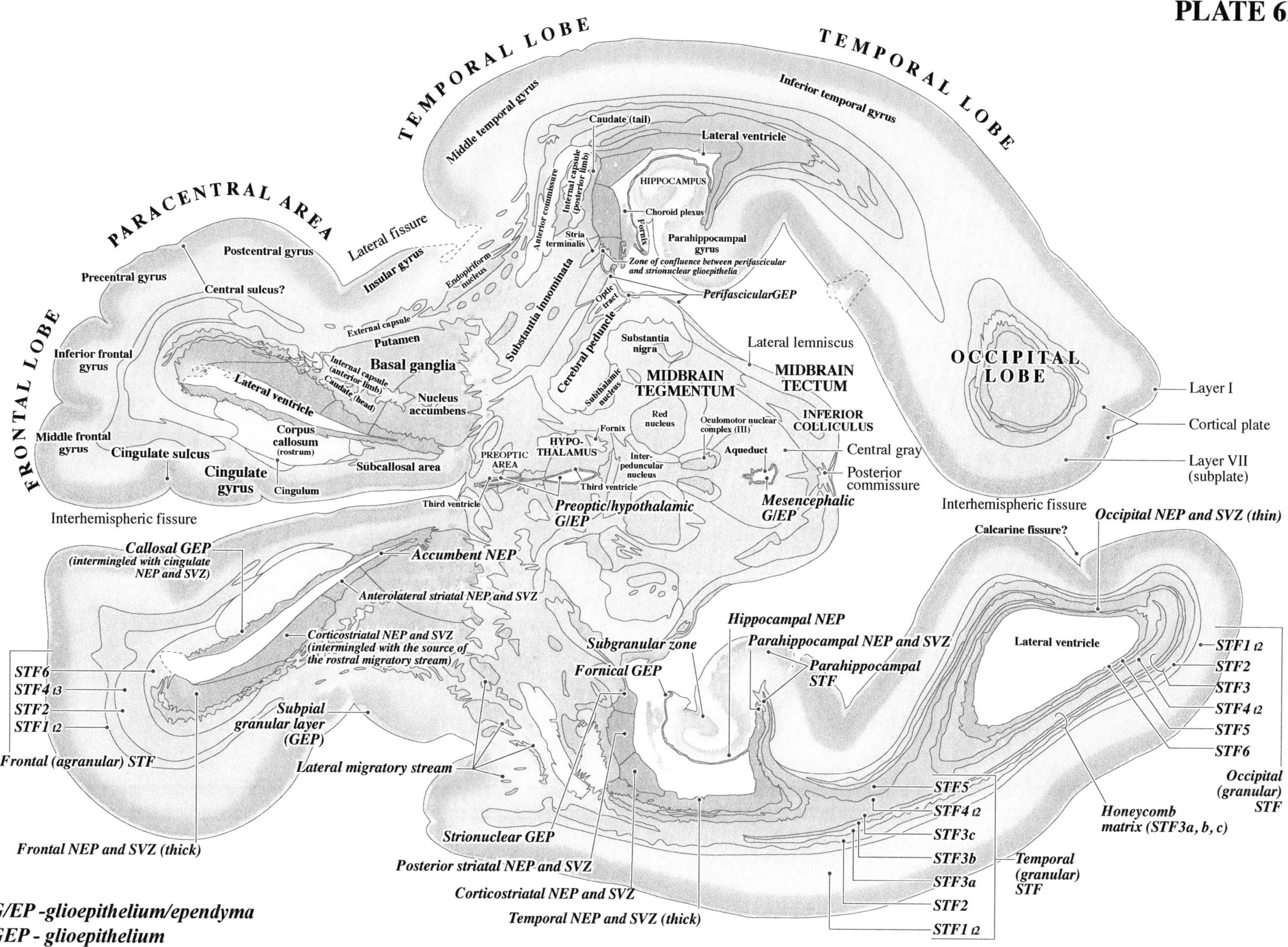

G/EP -glioepithelium/ependyma
GEP - glioepithelium
NEP - neuroepithelium
SVZ - subventricular zone
Germinal and transitional structures in *italics*

PLATE 7A
CR 190 mm
GW 23, Y197-65
Horizontal
Section 551

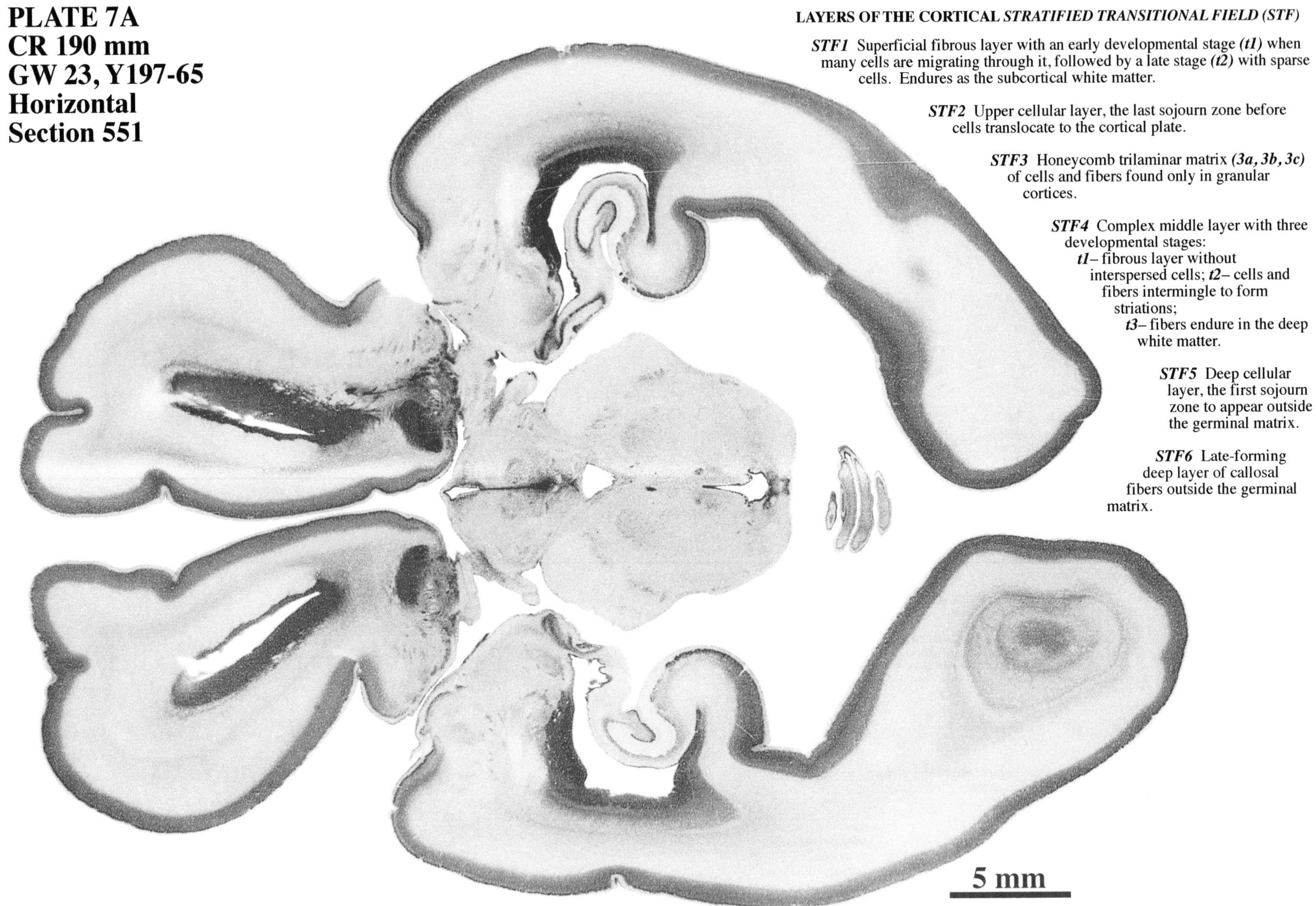

See detail of the brain core in Plates 14A and B.

PLATE 7B

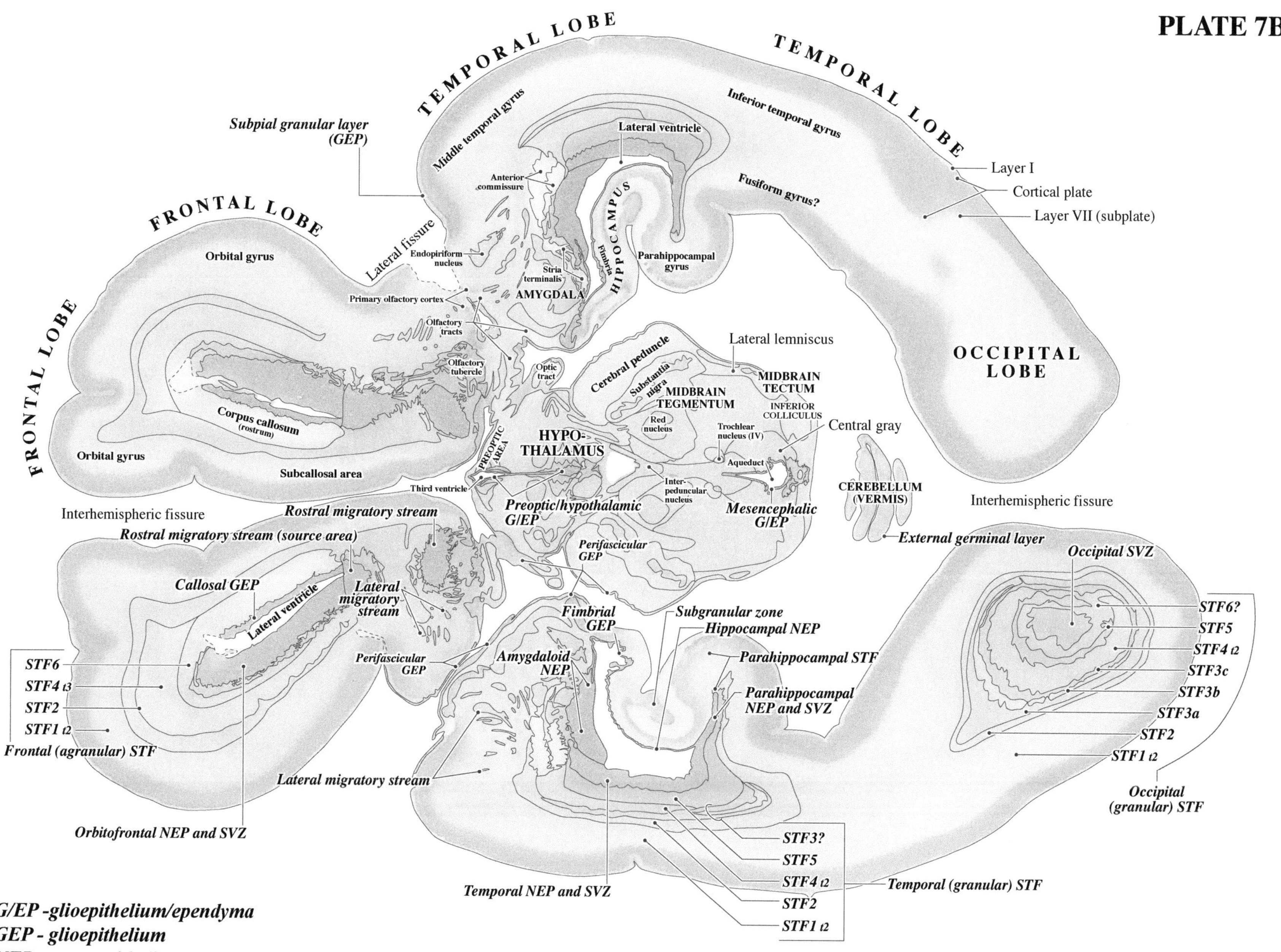

G/EP -glioepithelium/ependyma
GEP - glioepithelium
NEP - neuroepithelium
SVZ - subventricular zone
Germinal and transitional structures in *italics*

PLATE 8A
CR 190 mm
GW 23, Y197-65
Horizontal
Section 581

LAYERS OF THE CORTICAL *STRATIFIED TRANSITIONAL FIELD (STF)*

STF1 Superficial fibrous layer with an early developmental stage ***(t1)*** when many cells are migrating through it, followed by a late stage ***(t2)*** with sparse cells. Endures as the subcortical white matter.

STF2 Upper cellular layer, the last sojourn zone before cells translocate to the cortical plate.

STF3 Honeycomb trilaminar matrix ***(3a, 3b, 3c)*** of cells and fibers found only in granular cortices.

STF4 Complex middle layer with three developmental stages:
t1– fibrous layer without interspersed cells; ***t2***– cells and fibers intermingle to form striations;
t3– fibers endure in the deep white matter.

STF5 Deep cellular layer, the first sojourn zone to appear outside the germinal matrix.

STF6 Late-forming deep layer of callosal fibers outside the germinal matrix.

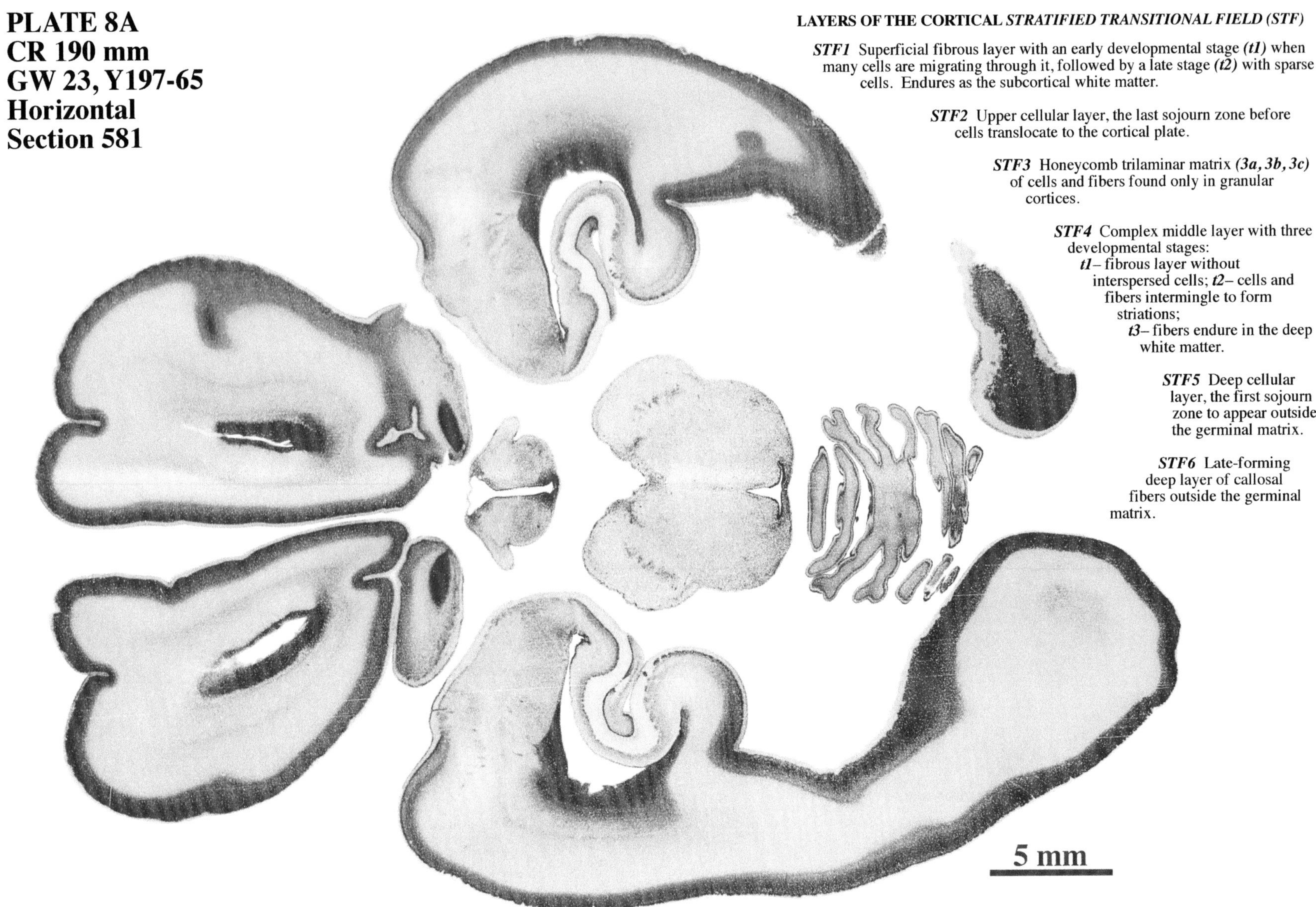

See detail of the brain core in Plates 15A and B.

PLATE 8B

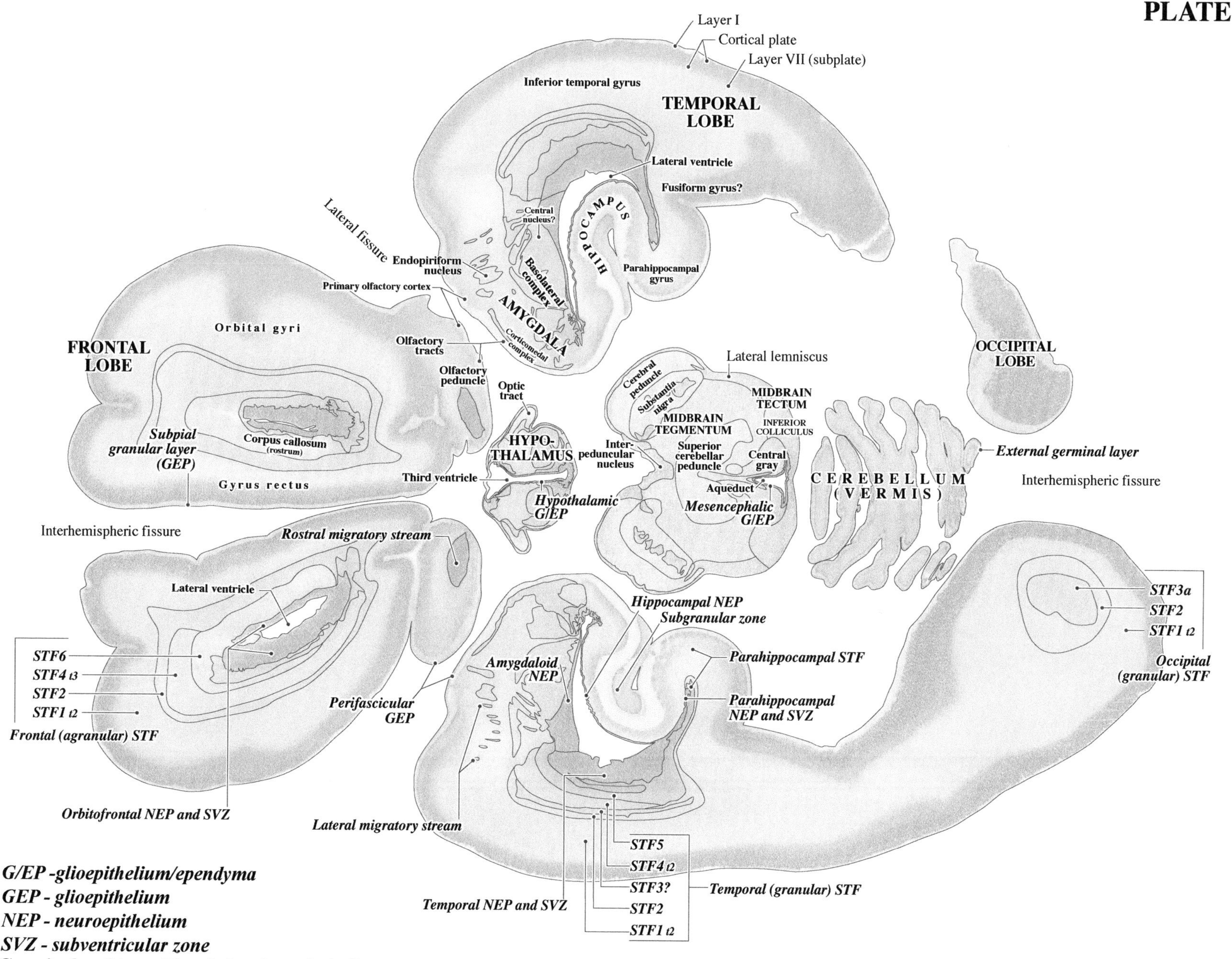

G/EP - glioepithelium/ependyma
GEP - glioepithelium
NEP - neuroepithelium
SVZ - subventricular zone
Germinal and transitional structures in *italics*

PLATE 9A
CR 190 mm
GW 23, Y197-65
Horizontal
Section 301

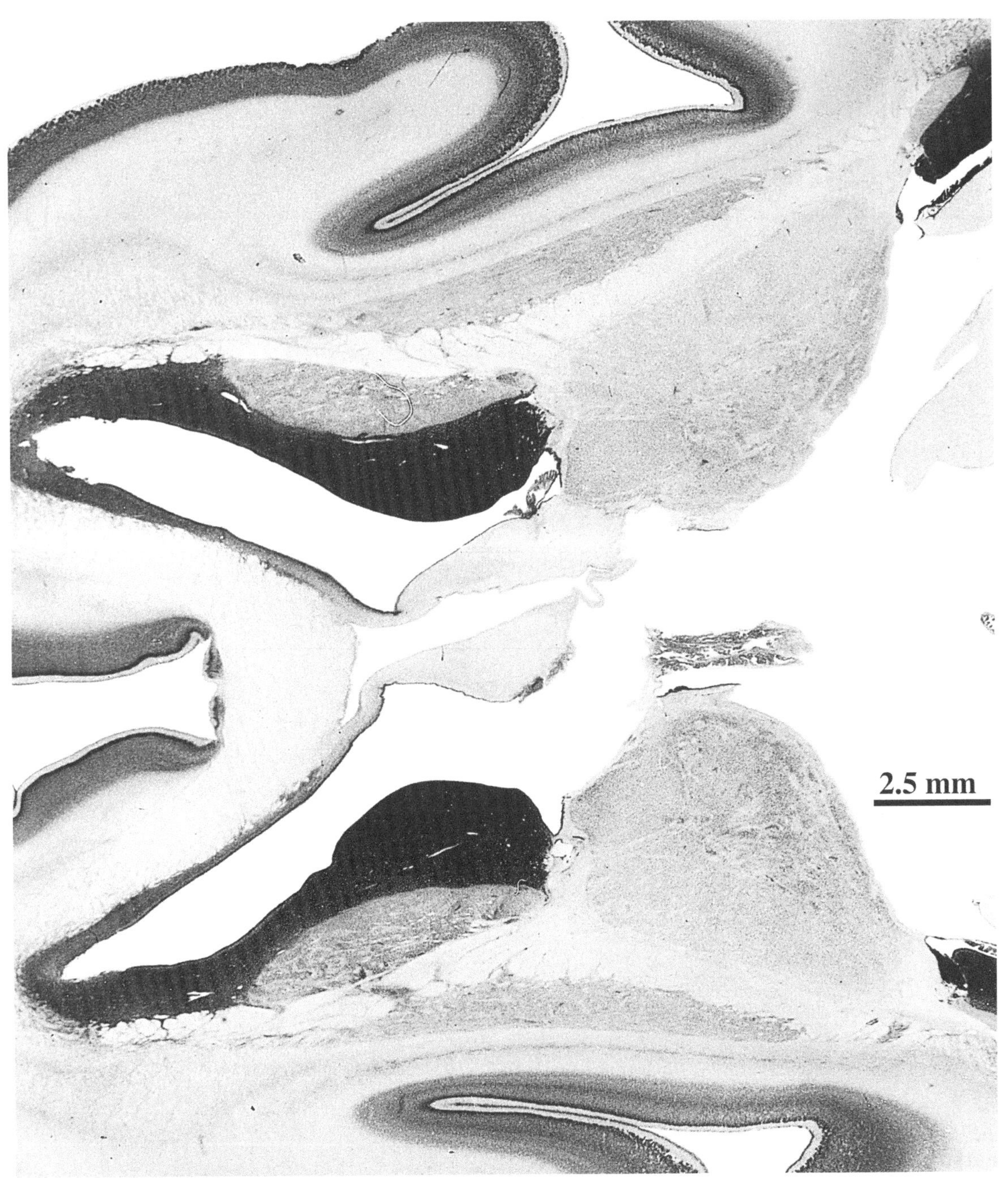

See the entire section in Plates 2A and B.

PLATE 9B

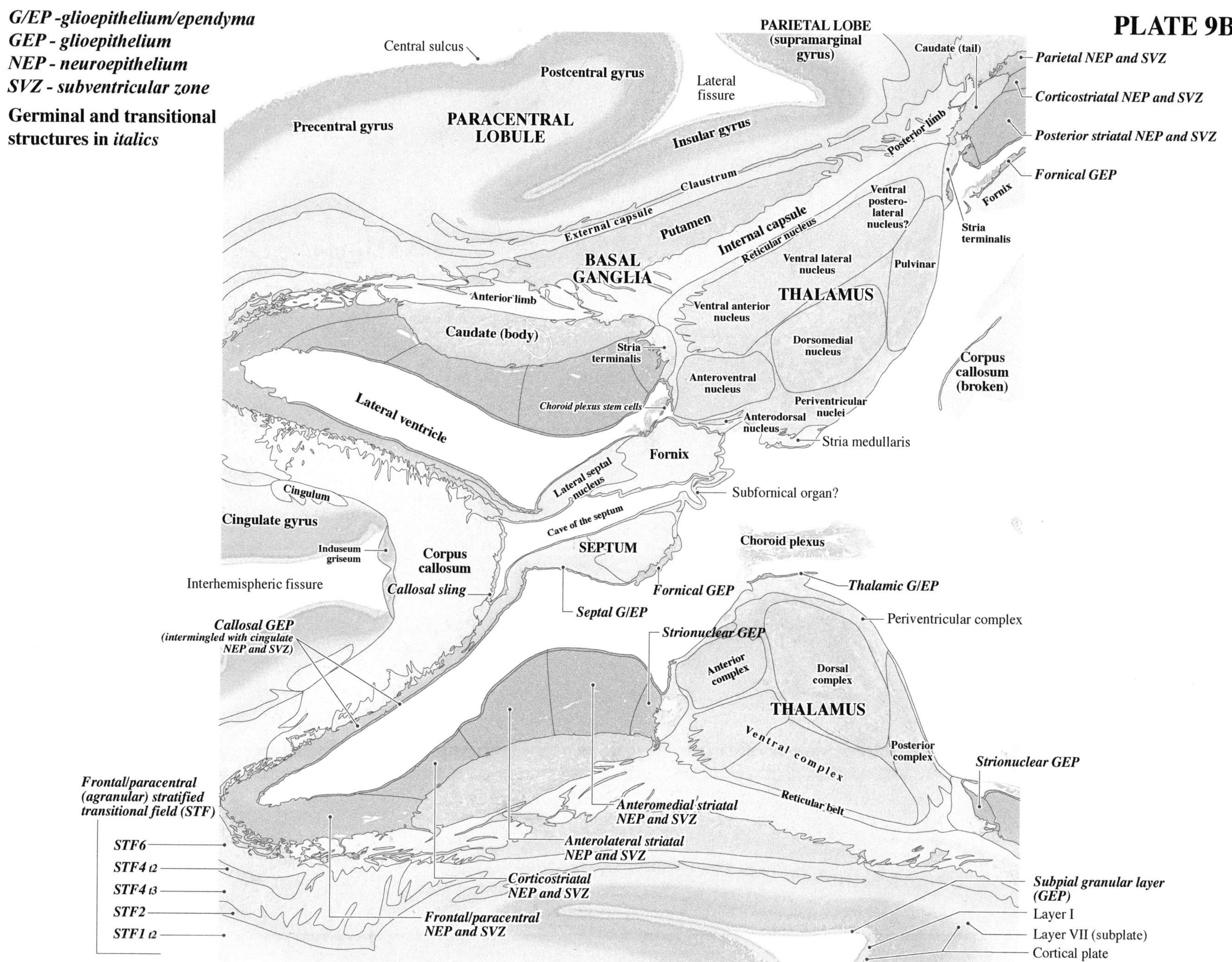

G/EP -glioepithelium/ependyma
GEP - glioepithelium
NEP - neuroepithelium
SVZ - subventricular zone
Germinal and transitional structures in *italics*
Central sulcus
Postcentral gyrus
PARIETAL LOBE (supramarginal gyrus)
Caudate (tail)
Parietal NEP and SVZ
Corticostriatal NEP and SVZ
Posterior striatal NEP and SVZ
Fornical GEP
Lateral fissure
Precentral gyrus
PARACENTRAL LOBULE
Insular gyrus
Posterior limb
Claustrum
External capsule
Putamen
Internal capsule
Reticular nucleus
Ventral postero-lateral nucleus?
Fornix
Stria terminalis
BASAL GANGLIA
Ventral lateral nucleus
Pulvinar
Anterior limb
THALAMUS
Ventral anterior nucleus
Caudate (body)
Dorsomedial nucleus
Stria terminalis
Corpus callosum (broken)
Anteroventral nucleus
Lateral ventricle
Choroid plexus stem cells
Periventricular nuclei
Anterodorsal nucleus
Stria medullaris
Fornix
Lateral septal nucleus
Subfornical organ?
Cingulum
Cingulate gyrus
Cave of the septum
SEPTUM
Choroid plexus
Induseum griseum
Corpus callosum
Interhemispheric fissure
Callosal sling
Fornical GEP
Thalamic G/EP
Septal G/EP
Periventricular complex
Callosal GEP (intermingled with cingulate NEP and SVZ)
Strionuclear GEP
Anterior complex
Dorsal complex
THALAMUS
Ventral complex
Posterior complex
Strionuclear GEP
Frontal/paracentral (agranular) stratified transitional field (STF)
Reticular belt
Anteromedial striatal NEP and SVZ
STF6
Anterolateral striatal NEP and SVZ
STF4 t2
STF4 t3
Corticostriatal NEP and SVZ
Subpial granular layer (GEP)
STF2
Layer I
Frontal/paracentral NEP and SVZ
STF1 t2
Layer VII (subplate)
Cortical plate

PLATE 10A
CR 190 mm
GW 23, Y197-65
Horizontal
Section 361

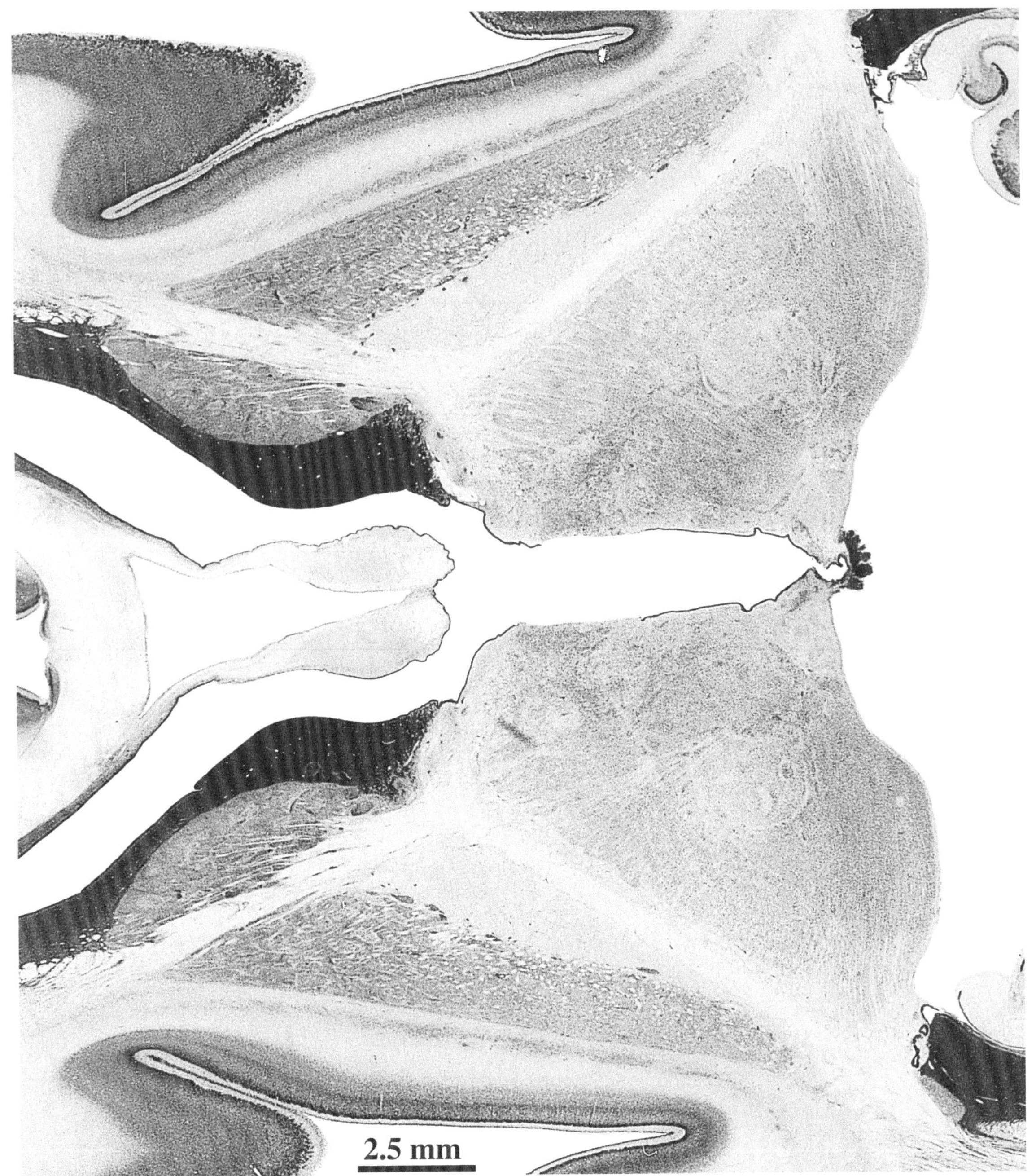

See the entire section in Plates 3A and B.

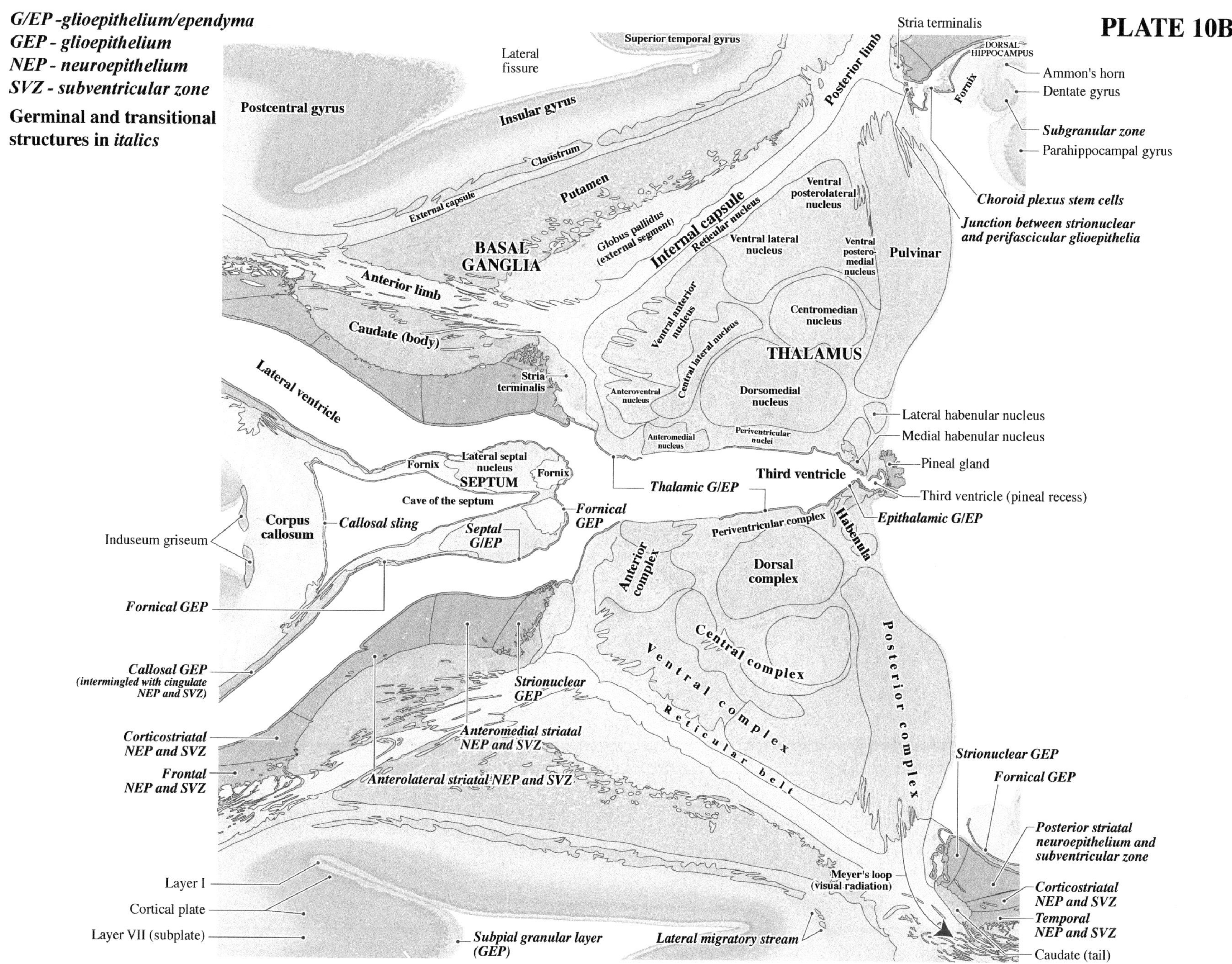
PLATE 10B
G/EP -glioepithelium/ependyma
GEP - glioepithelium
NEP - neuroepithelium
SVZ - subventricular zone
Germinal and transitional structures in italics
Postcentral gyrus
Lateral fissure
Superior temporal gyrus
Insular gyrus
Claustrum
External capsule
Putamen
BASAL GANGLIA
Globus pallidus (external segment)
Internal capsule
Reticular nucleus
Posterior limb
Stria terminalis
DORSAL HIPPOCAMPUS
Fornix
Ammon's horn
Dentate gyrus
Subgranular zone
Parahippocampal gyrus
Choroid plexus stem cells
Junction between strionuclear and perifascicular glioepithelia
Ventral posterolateral nucleus
Ventral lateral nucleus
Ventral postero-medial nucleus
Pulvinar
Anterior limb
Caudate (body)
Ventral anterior nucleus
Central lateral nucleus
Centromedian nucleus
THALAMUS
Stria terminalis
Lateral ventricle
Anteroventral nucleus
Dorsomedial nucleus
Anteromedial nucleus
Periventricular nuclei
Lateral habenular nucleus
Medial habenular nucleus
Pineal gland
Fornix
Lateral septal nucleus
SEPTUM
Fornix
Third ventricle
Thalamic G/EP
Third ventricle (pineal recess)
Cave of the septum
Fornical GEP
Epithalamic G/EP
Corpus callosum
Callosal sling
Septal G/EP
Periventricular complex
Habenula
Induseum griseum
Anterior complex
Dorsal complex
Fornical GEP
Central complex
Posterior complex
Ventral complex
Callosal GEP (intermingled with cingulate NEP and SVZ)
Strionuclear GEP
Reticular belt
Corticostriatal NEP and SVZ
Anteromedial striatal NEP and SVZ
Strionuclear GEP
Frontal NEP and SVZ
Anterolateral striatal NEP and SVZ
Fornical GEP
Posterior striatal neuroepithelium and subventricular zone
Layer I
Meyer's loop (visual radiation)
Corticostriatal NEP and SVZ
Cortical plate
Temporal NEP and SVZ
Layer VII (subplate)
Subpial granular layer (GEP)
Lateral migratory stream
Caudate (tail)

PLATE 11A
CR 190 mm
GW 23, Y197-65
Horizontal
Section 431

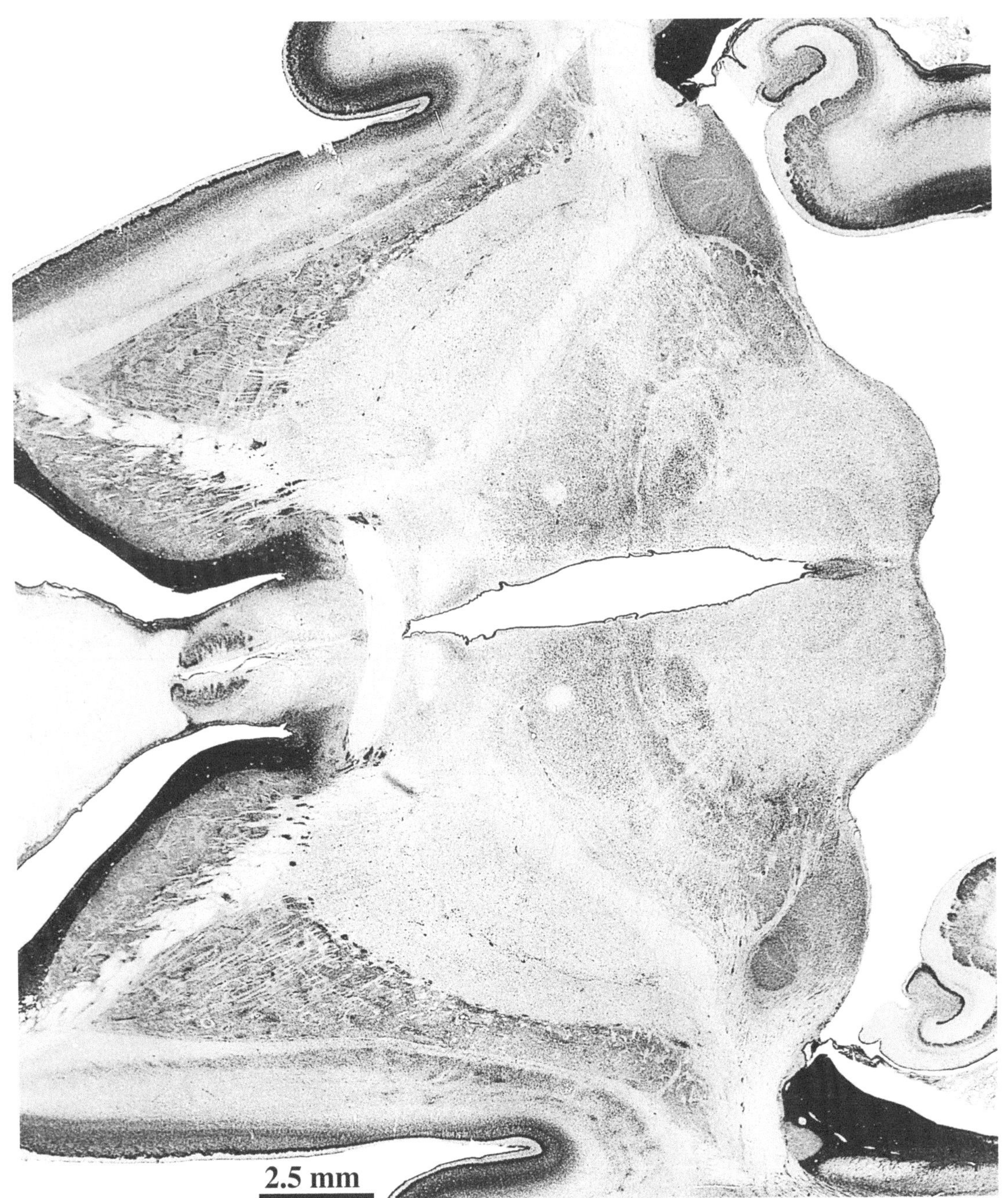

See the entire section in Plates 4A and B.

PLATE 11B

G/EP -glioepithelium/ependyma
GEP - glioepithelium
NEP - neuroepithelium
SVZ - subventricular zone

Germinal and transitional structures in *italics*

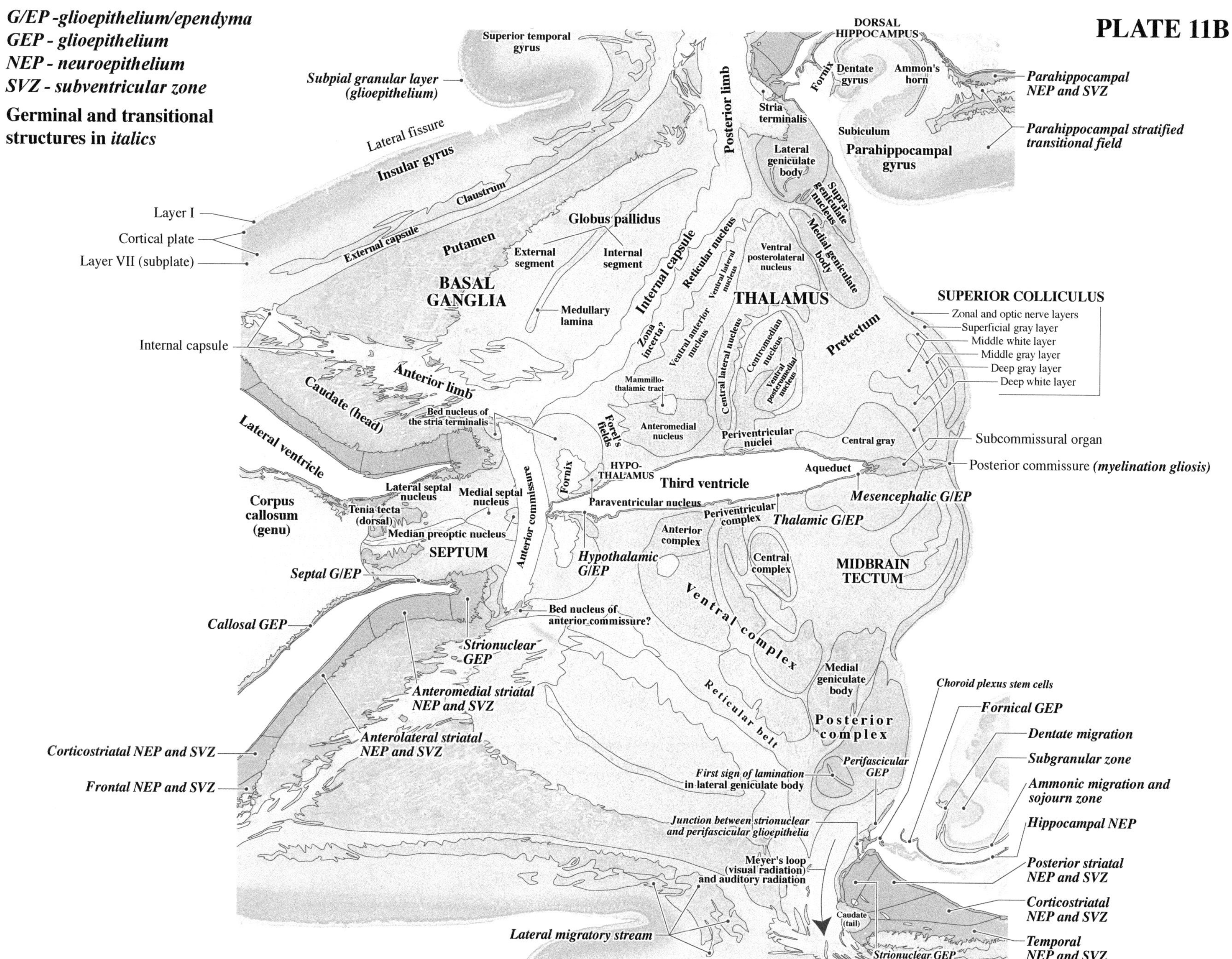

PLATE 12A
CR 190 mm
GW 23, Y197-65
Horizontal
Section 491

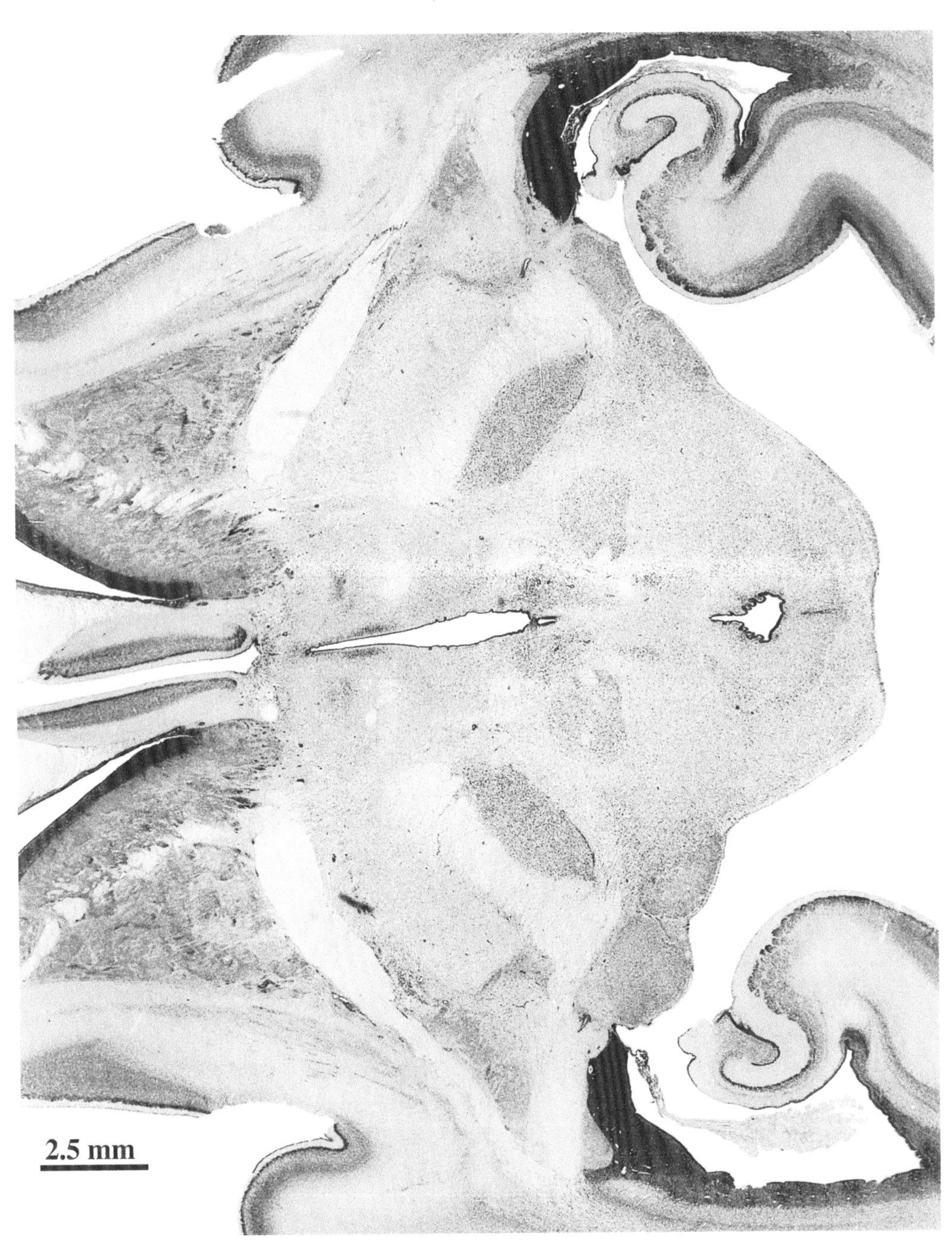

See the entire section in Plates 5A and B.

G/EP -glioepithelium/ependyma
GEP - glioepithelium
NEP - neuroepithelium
SVZ - subventricular zone

Germinal and transitional structures in *italics*

PLATE 12B

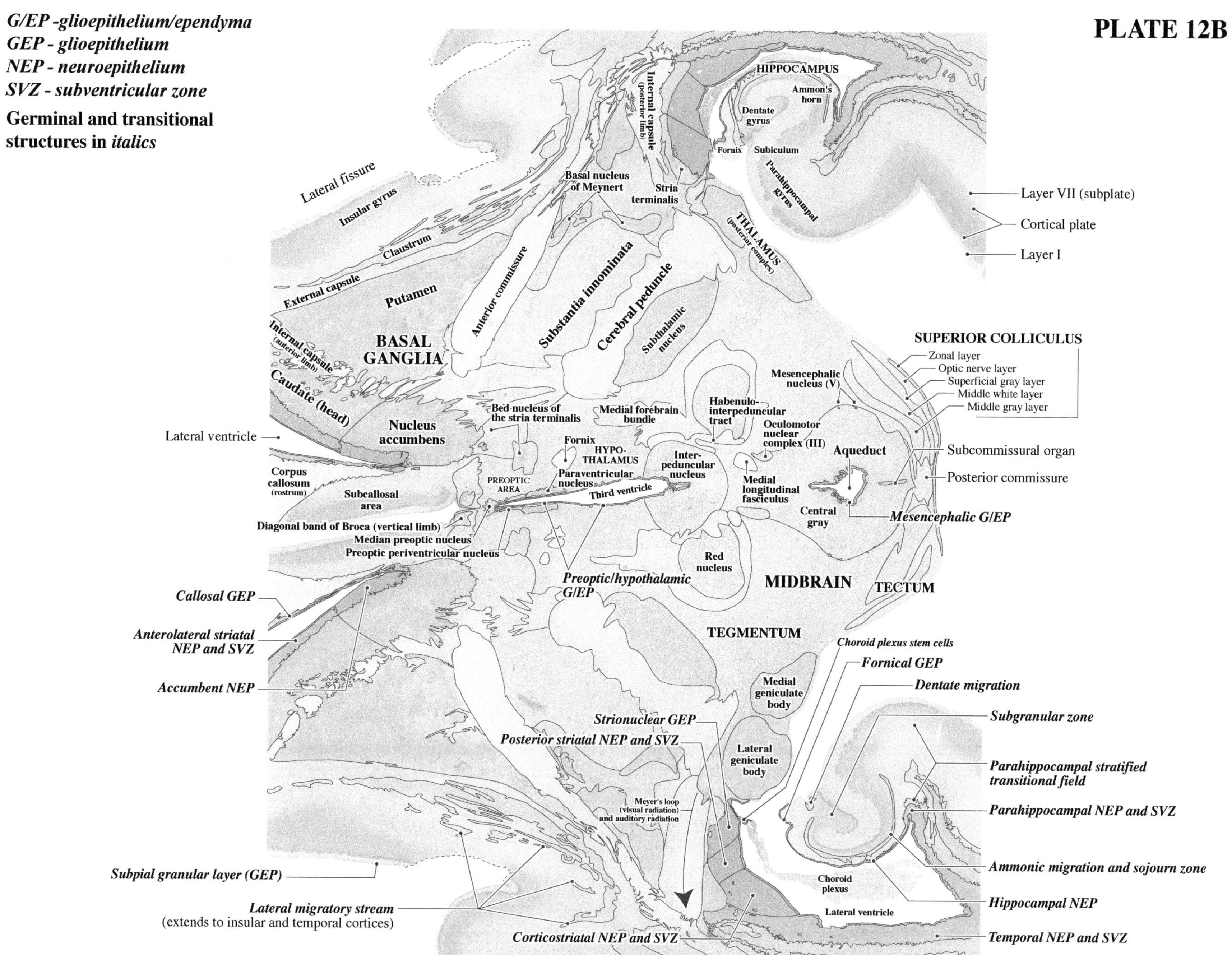

PLATE 13A
CR 190 mm
GW 23, Y197-65
Horizontal
Section 521

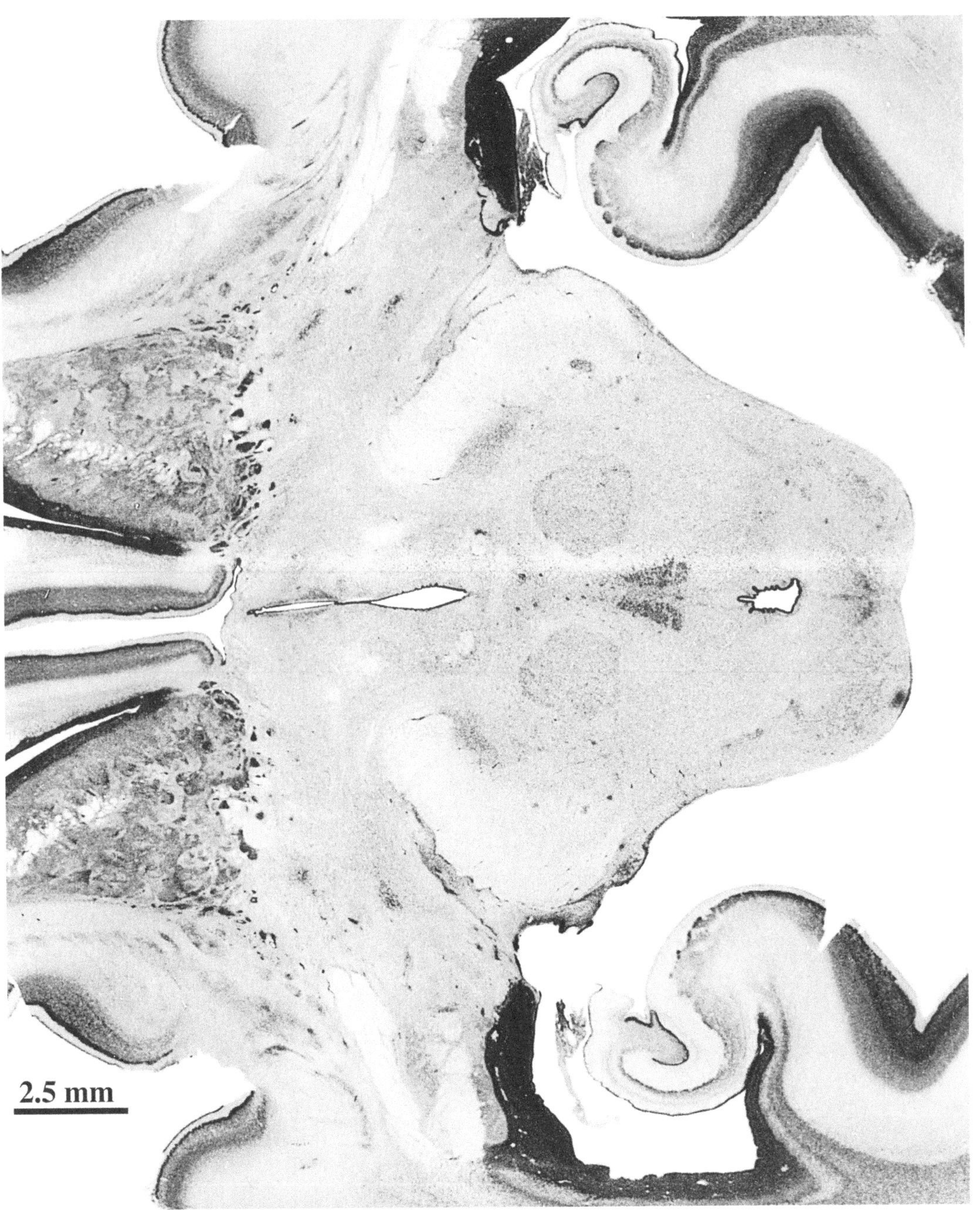

See the entire section in Plates 6A and B.

PLATE 13B

G/EP -glioepithelium/ependyma
GEP - glioepithelium
NEP - neuroepithelium
SVZ - subventricular zone

Germinal and transitional structures in *italics*

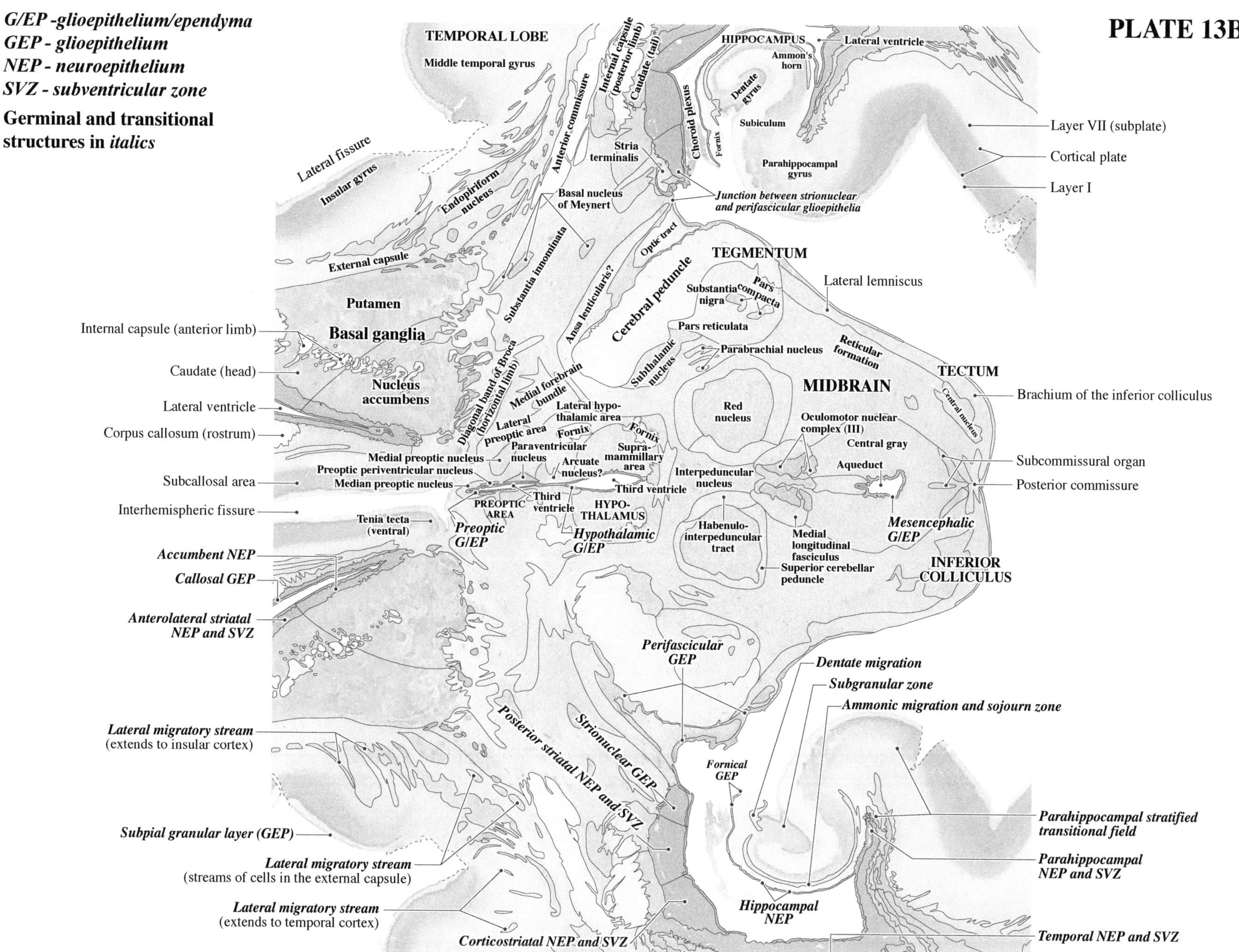

PLATE 14A
CR 190 mm
GW 23, Y197-65
Horizontal
Section 551

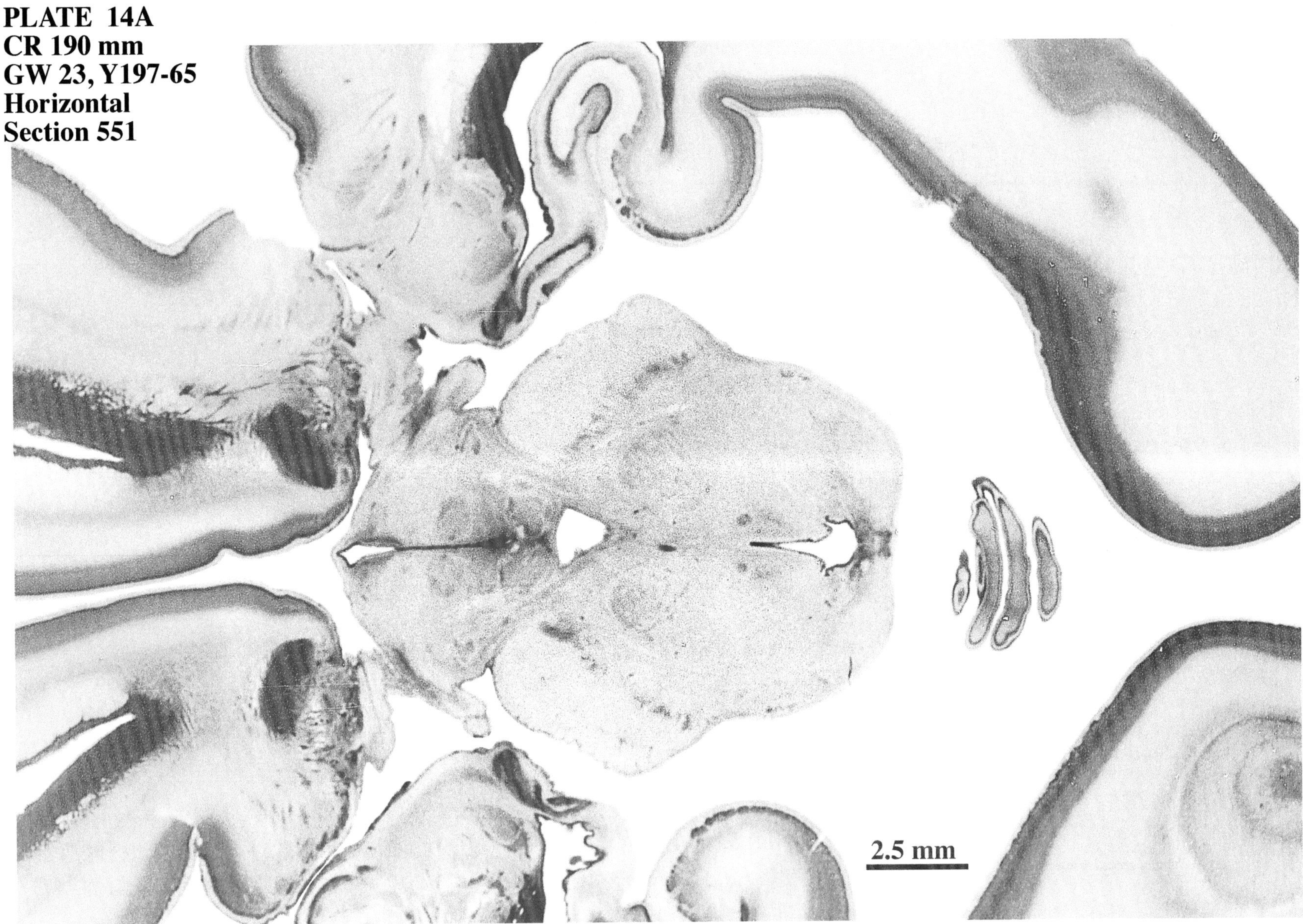

See the entire section in Plates 7A and B.

PLATE 14B

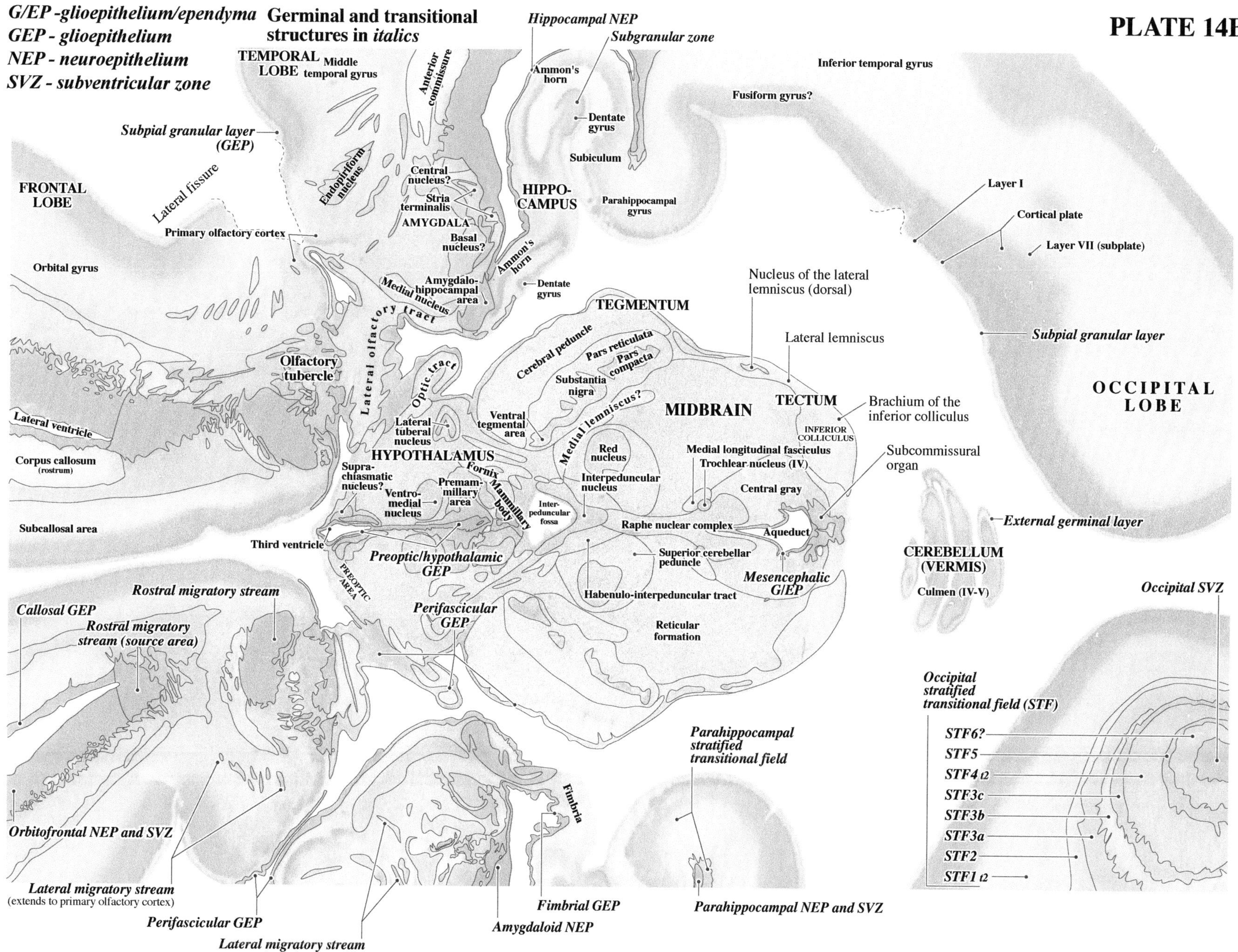
G/EP - glioepithelium/ependyma
GEP - glioepithelium
NEP - neuroepithelium
SVZ - subventricular zone
Germinal and transitional structures in italics
TEMPORAL LOBE
Middle temporal gyrus
Anterior commissure
Hippocampal NEP
Subgranular zone
Ammon's horn
Dentate gyrus
Subiculum
Inferior temporal gyrus
Fusiform gyrus?
Subpial granular layer (GEP)
Endopiriform nucleus
Central nucleus?
Stria terminalis
AMYGDALA
Basal nucleus?
HIPPOCAMPUS
Parahippocampal gyrus
FRONTAL LOBE
Lateral fissure
Primary olfactory cortex
Orbital gyrus
Medial nucleus
Amygdalo-hippocampal area
Ammon's horn
Dentate gyrus
Layer I
Cortical plate
Layer VII (subplate)
Nucleus of the lateral lemniscus (dorsal)
TEGMENTUM
Lateral olfactory tract
Cerebral peduncle
Pars reticulata
Pars compacta
Lateral lemniscus
Subpial granular layer
Olfactory tubercle
Optic tract
Substantia nigra
OCCIPITAL LOBE
Medial lemniscus?
MIDBRAIN
TECTUM
Brachium of the inferior colliculus
Lateral ventricle
Lateral tuberal nucleus
Ventral tegmental area
INFERIOR COLLICULUS
HYPOTHALAMUS
Red nucleus
Medial longitudinal fasciculus
Subcommissural organ
Corpus callosum (rostrum)
Fornix
Supra-chiasmatic nucleus?
Premam-millary area
Mammillary body
Interpeduncular nucleus
Trochlear nucleus (IV)
Central gray
Ventro-medial nucleus
Inter-peduncular fossa
External germinal layer
Subcallosal area
Raphe nuclear complex
Aqueduct
Third ventricle
Preoptic/hypothalamic GEP
Superior cerebellar peduncle
CEREBELLUM (VERMIS)
PREOPTIC AREA
Mesencephalic G/EP
Culmen (IV-V)
Occipital SVZ
Rostral migratory stream
Habenulo-interpeduncular tract
Callosal GEP
Rostral migratory stream (source area)
Perifascicular GEP
Reticular formation
Occipital stratified transitional field (STF)
STF6?
STF5
STF4 t2
STF3c
STF3b
STF3a
STF2
STF1 t2
Parahippocampal stratified transitional field
Fimbria
Orbitofrontal NEP and SVZ
Lateral migratory stream (extends to primary olfactory cortex)
Perifascicular GEP
Fimbrial GEP
Parahippocampal NEP and SVZ
Amygdaloid NEP
Lateral migratory stream (extends to amygdala)

PLATE 15A
CR 190 mm
GW 23, Y197-65
Horizontal
Section 581

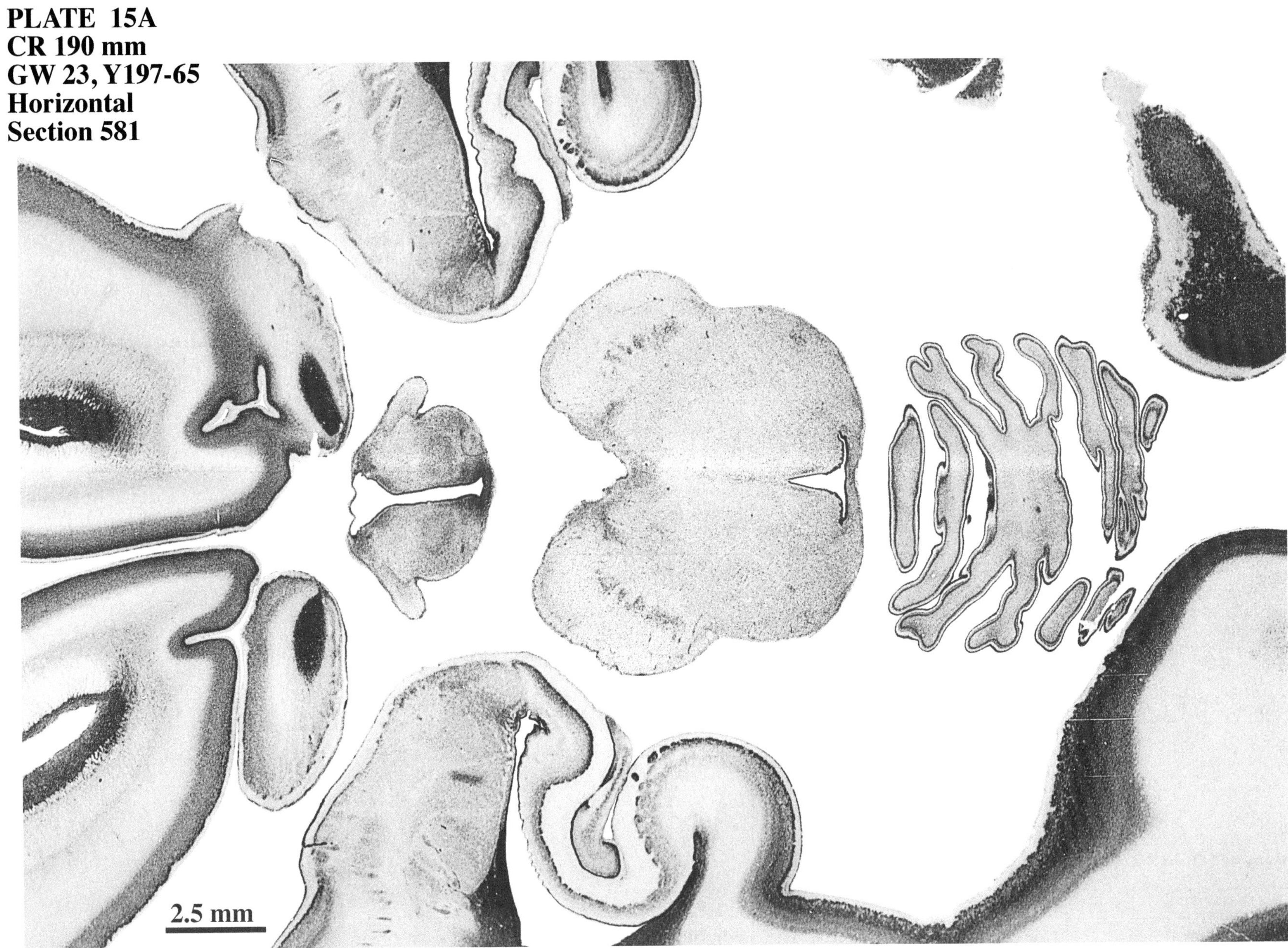

See the entire section in Plates 8A and B.

PLATE 15B

G/EP - glioepithelium/ependyma
GEP - glioepithelium
NEP - neuroepithelium
SVZ - subventricular zone

Germinal and transitional structures in *italics*

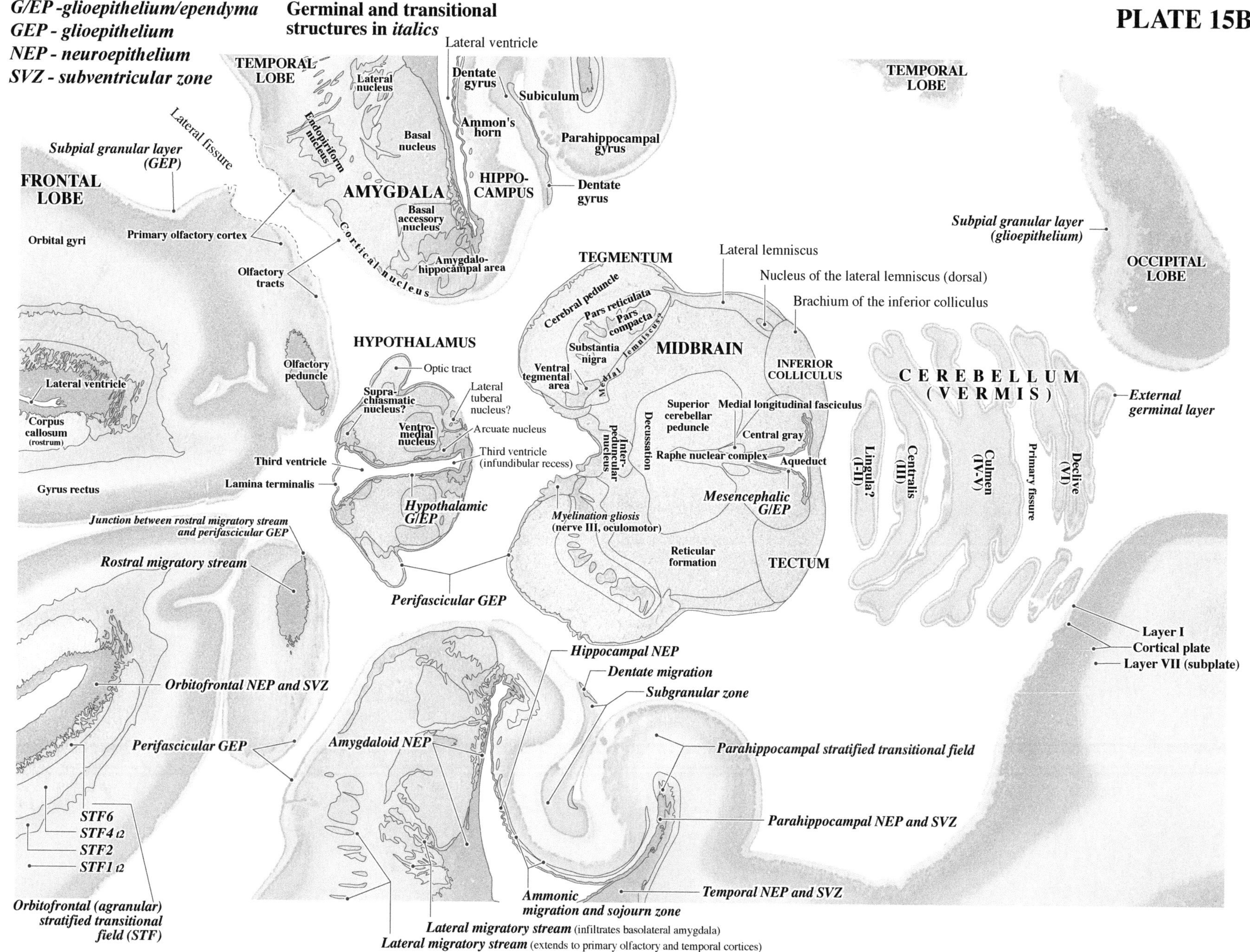

PLATE 16A
CR 190 mm
GW 23, Y197-65
Horizontal
Section 641

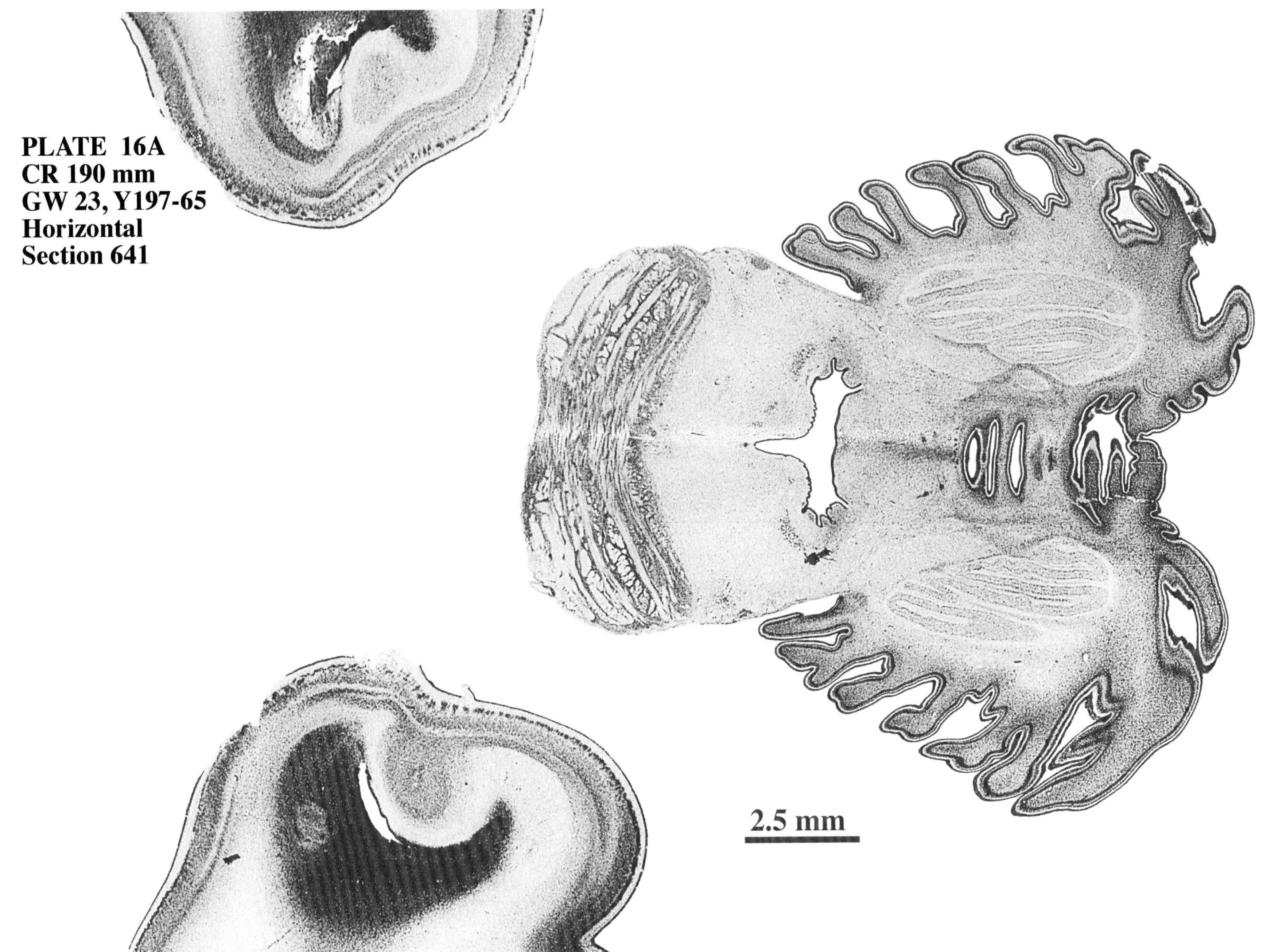

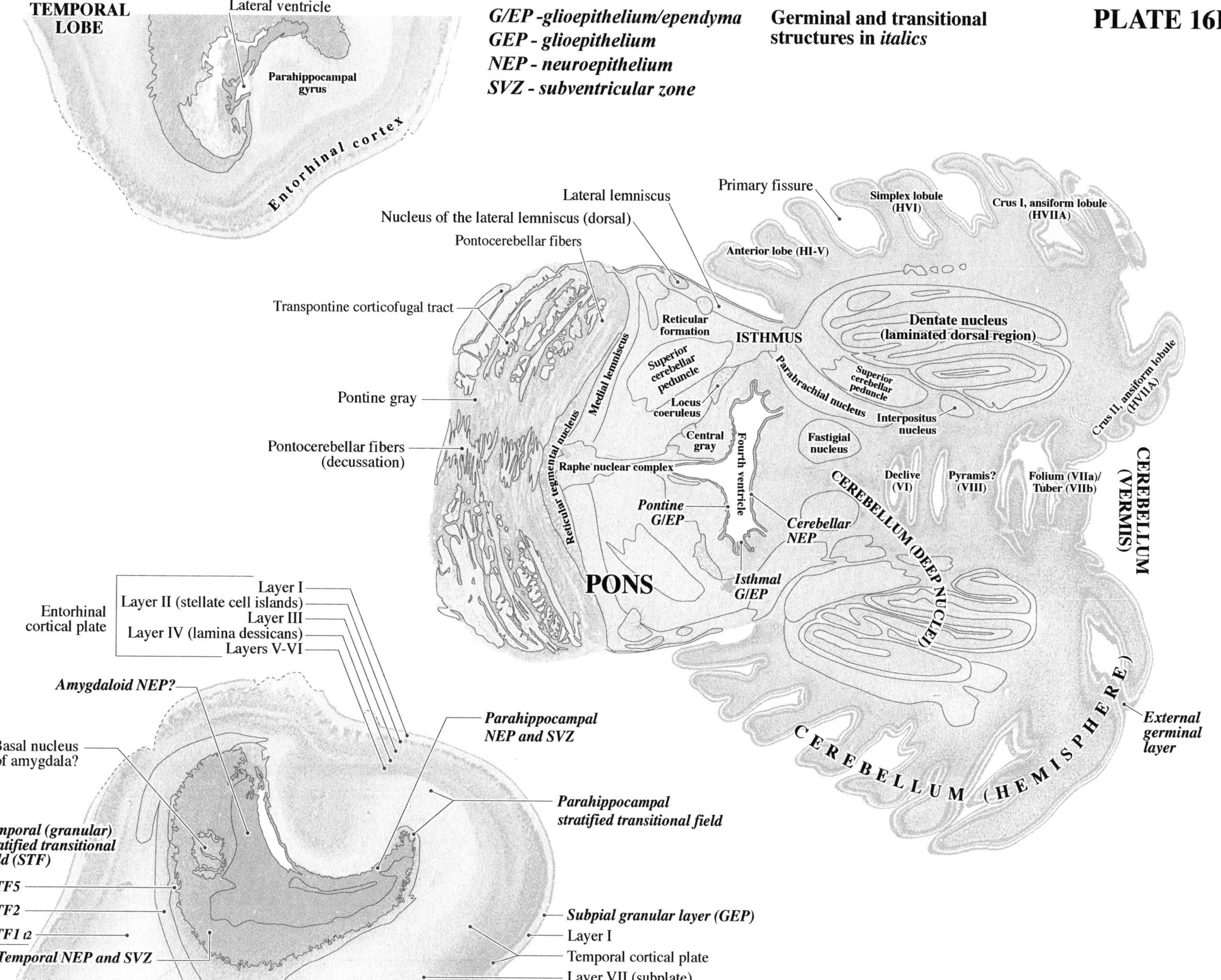

PLATE 16B
G/EP -glioepithelium/ependyma
GEP - glioepithelium
NEP - neuroepithelium
SVZ - subventricular zone
Germinal and transitional structures in italics
TEMPORAL LOBE
Lateral ventricle
Parahippocampal gyrus
Entorhinal cortex
Lateral lemniscus
Nucleus of the lateral lemniscus (dorsal)
Pontocerebellar fibers
Transpontine corticofugal tract
Pontine gray
Pontocerebellar fibers (decussation)
Primary fissure
Simplex lobule (HVI)
Crus I, ansiform lobule (HVIIA)
Anterior lobe (HI-V)
Dentate nucleus (laminated dorsal region)
Reticular formation
ISTHMUS
Medial lemniscus
Superior cerebellar peduncle
Locus coeruleus
Parabrachial nucleus
Superior cerebellar peduncle
Interpositus nucleus
Crus II, ansiform lobule (HVIIA)
Central gray
Fastigial nucleus
Reticular tegmental nucleus
Raphe nuclear complex
Fourth ventricle
Declive (VI)
Pyramis? (VIII)
Folium (VIIa)/ Tuber (VIIb)
CEREBELLUM (VERMIS)
CEREBELLUM (DEEP NUCLEI)
Pontine G/EP
Cerebellar NEP
Isthmal G/EP
PONS
Entorhinal cortical plate
Layer I
Layer II (stellate cell islands)
Layer III
Layer IV (lamina dessicans)
Layers V-VI
Amygdaloid NEP?
Basal nucleus of amygdala?
Parahippocampal NEP and SVZ
Parahippocampal stratified transitional field
CEREBELLUM (HEMISPHERE)
External germinal layer
Temporal (granular) stratified transitional field (STF)
STF5
STF2
STF1 t2
Temporal NEP and SVZ
Subpial granular layer (GEP)
Layer I
Temporal cortical plate
Layer VII (subplate)

PLATE 17A
CR 190 mm
GW 23, Y197-65
Horizontal
Section 691

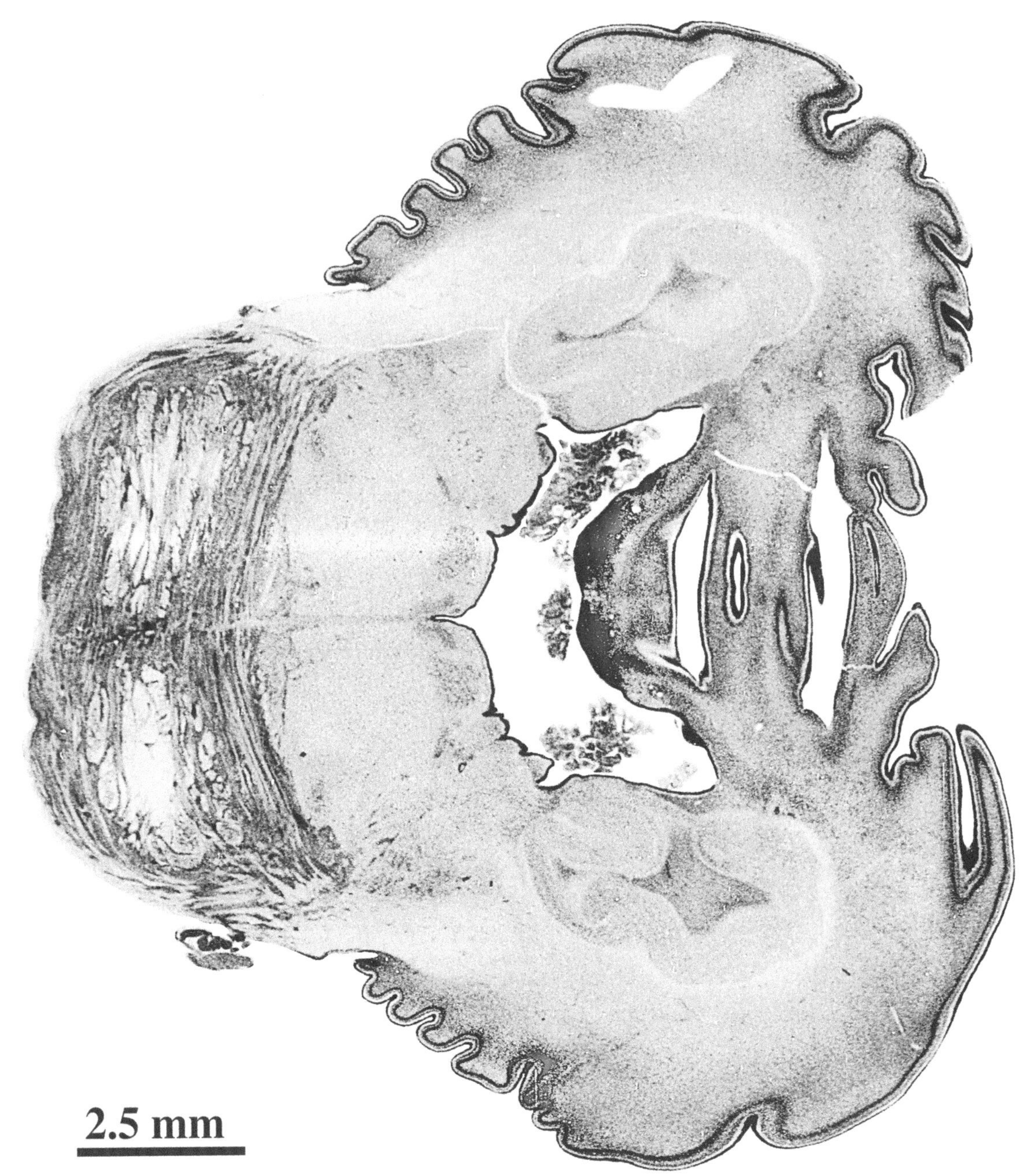

PLATE 17B

G/EP -glioepithelium/ependyma
NEP - neuroepithelium

Germinal and transitional structures in *italics*

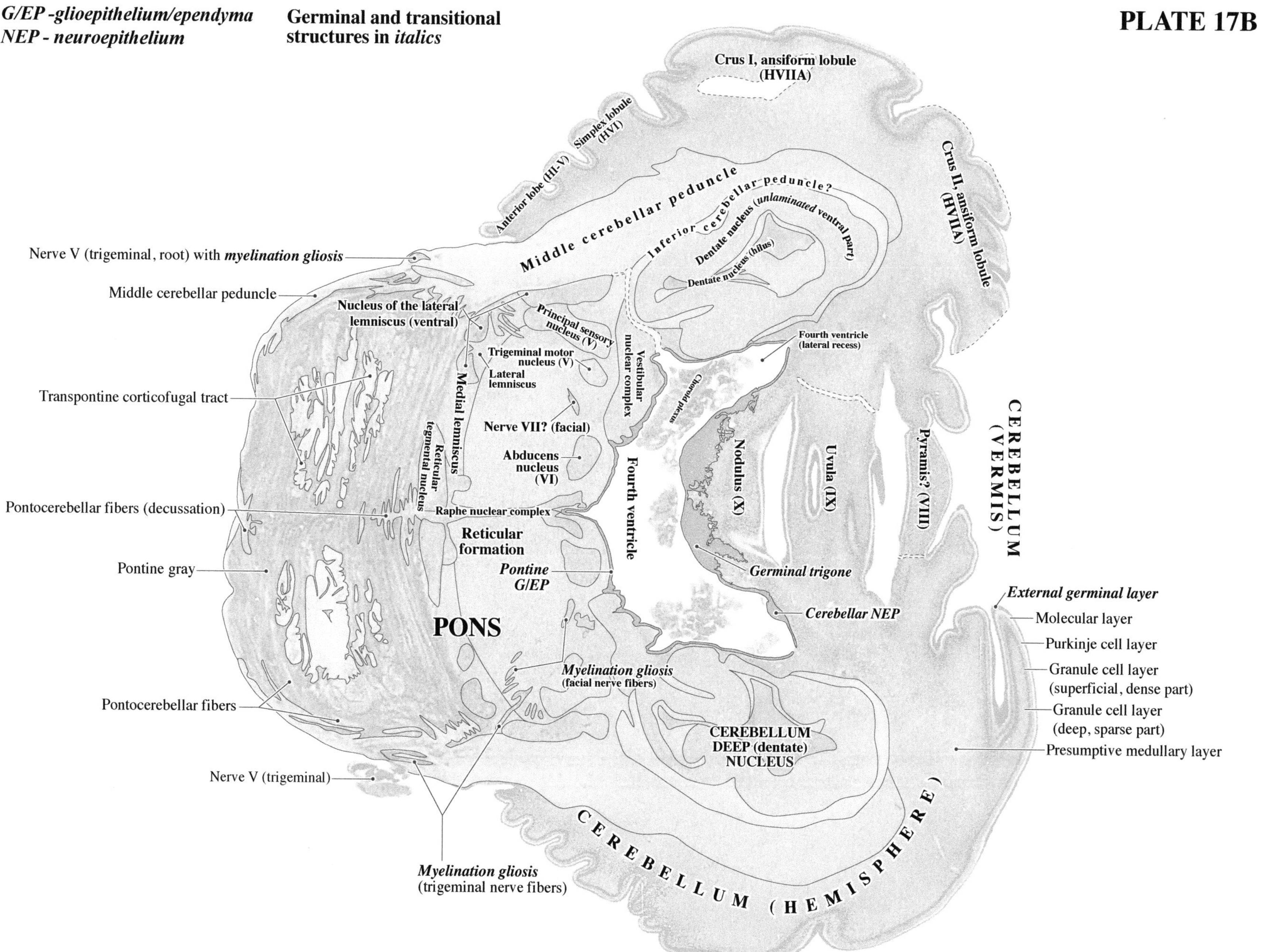

PLATE 18A
CR 190 mm
GW 23, Y197-65
Horizontal
Section 711

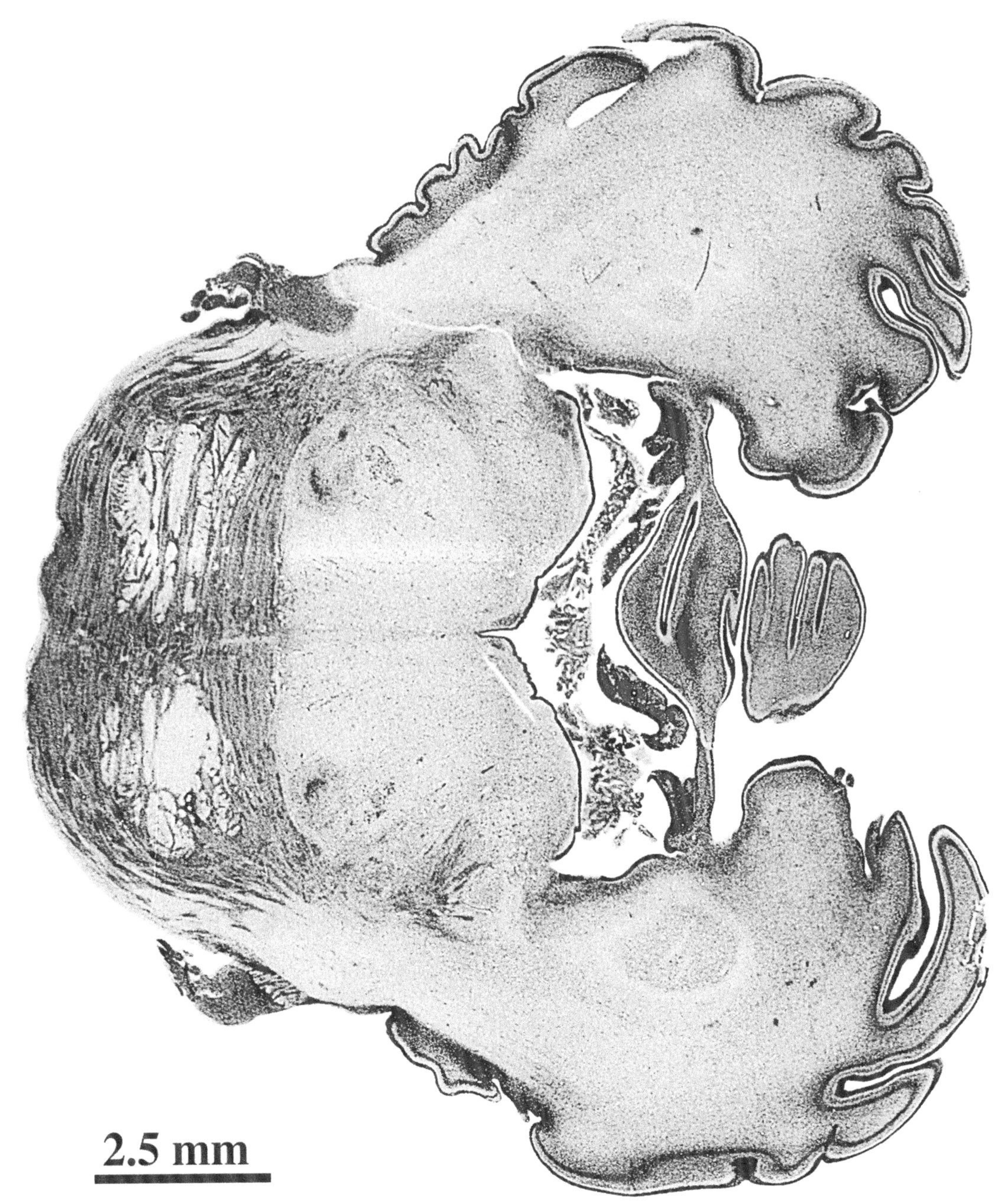

G/EP -glioepithelium/ependyma

Germinal and transitional structures in *italics*

PLATE 18B

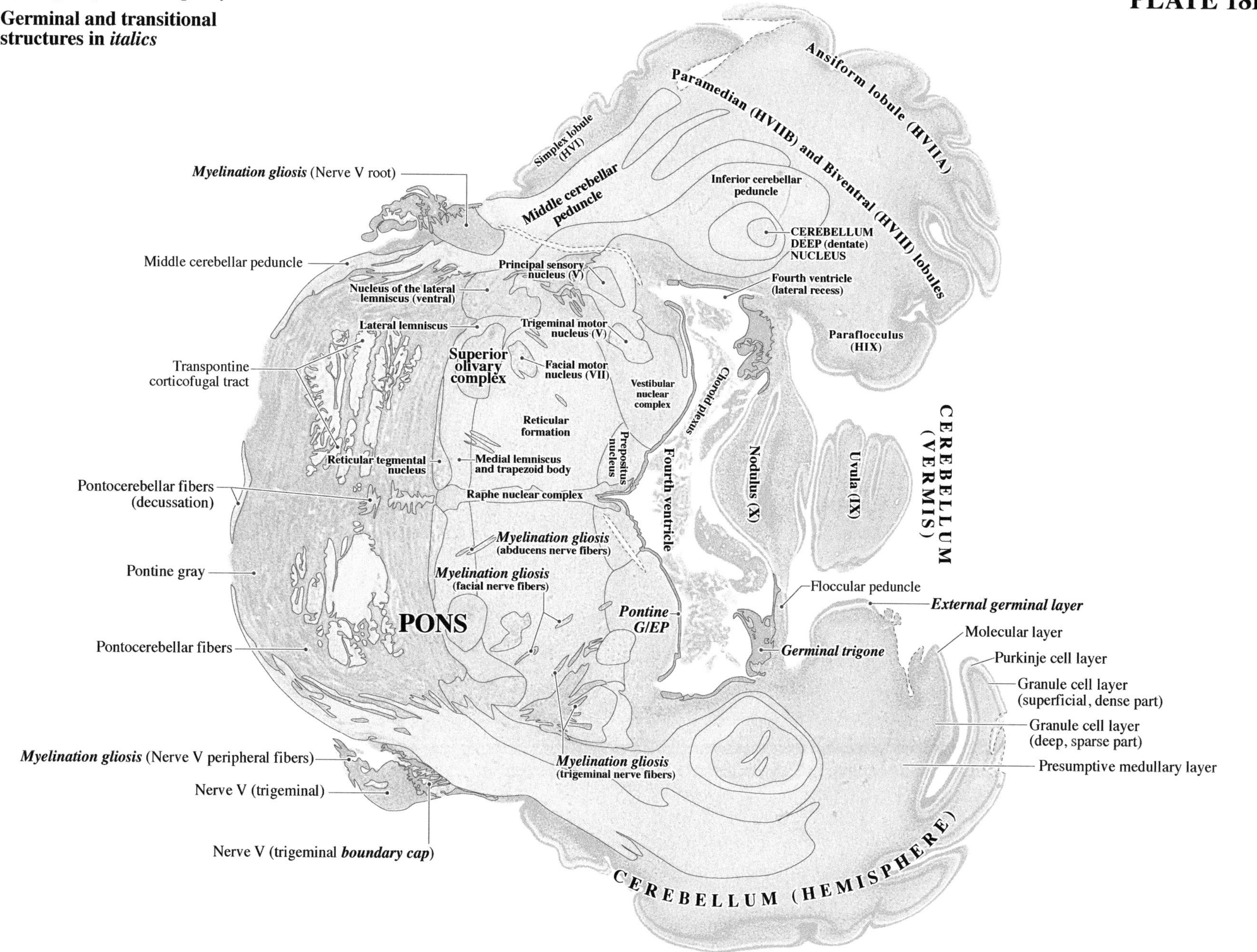

PLATE 19A
CR 190 mm
GW 23, Y197-65
Horizontal
Section 741

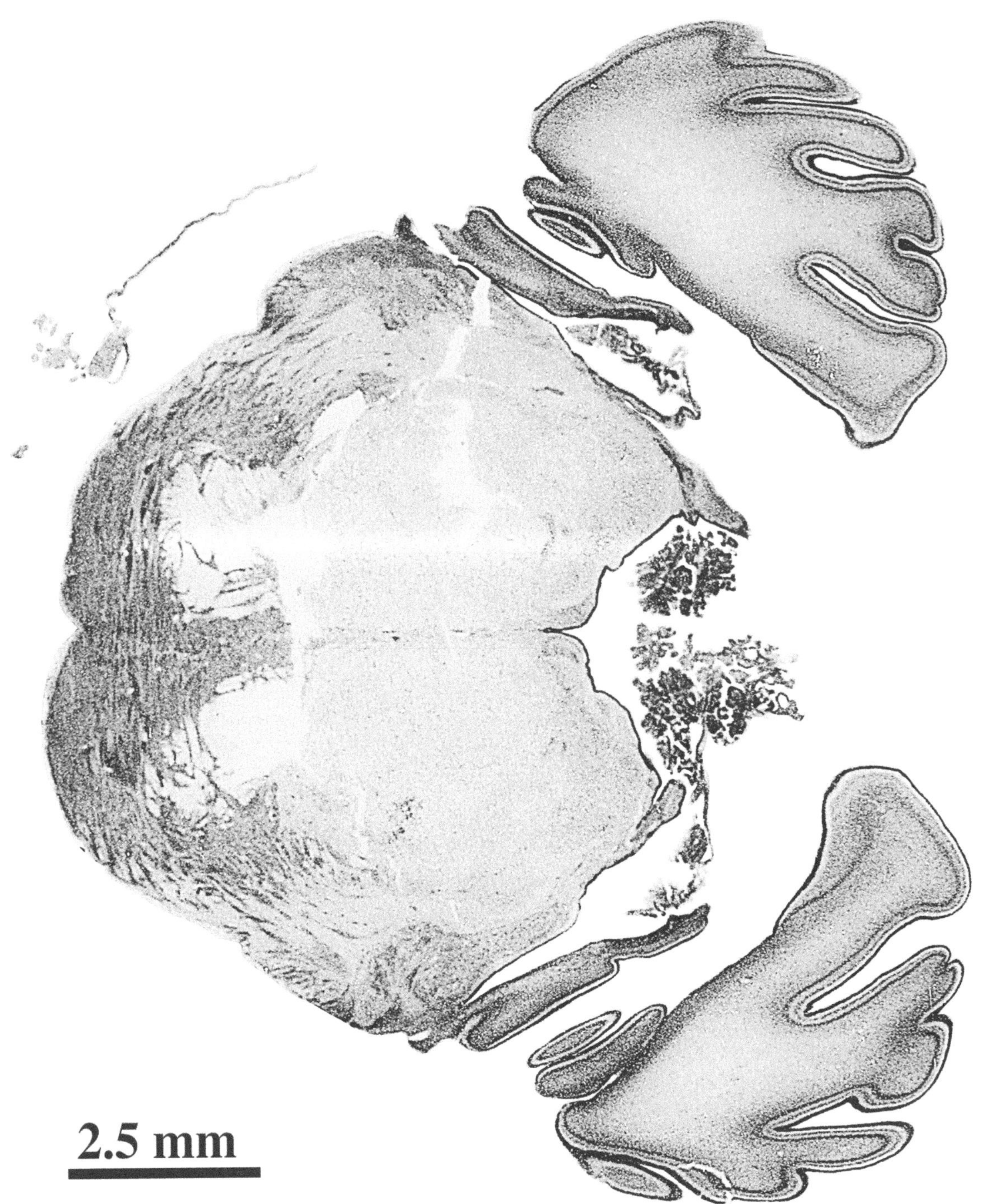

PLATE 19B

G/EP -glioepithelium/ependyma

Germinal and transitional structures in *italics*

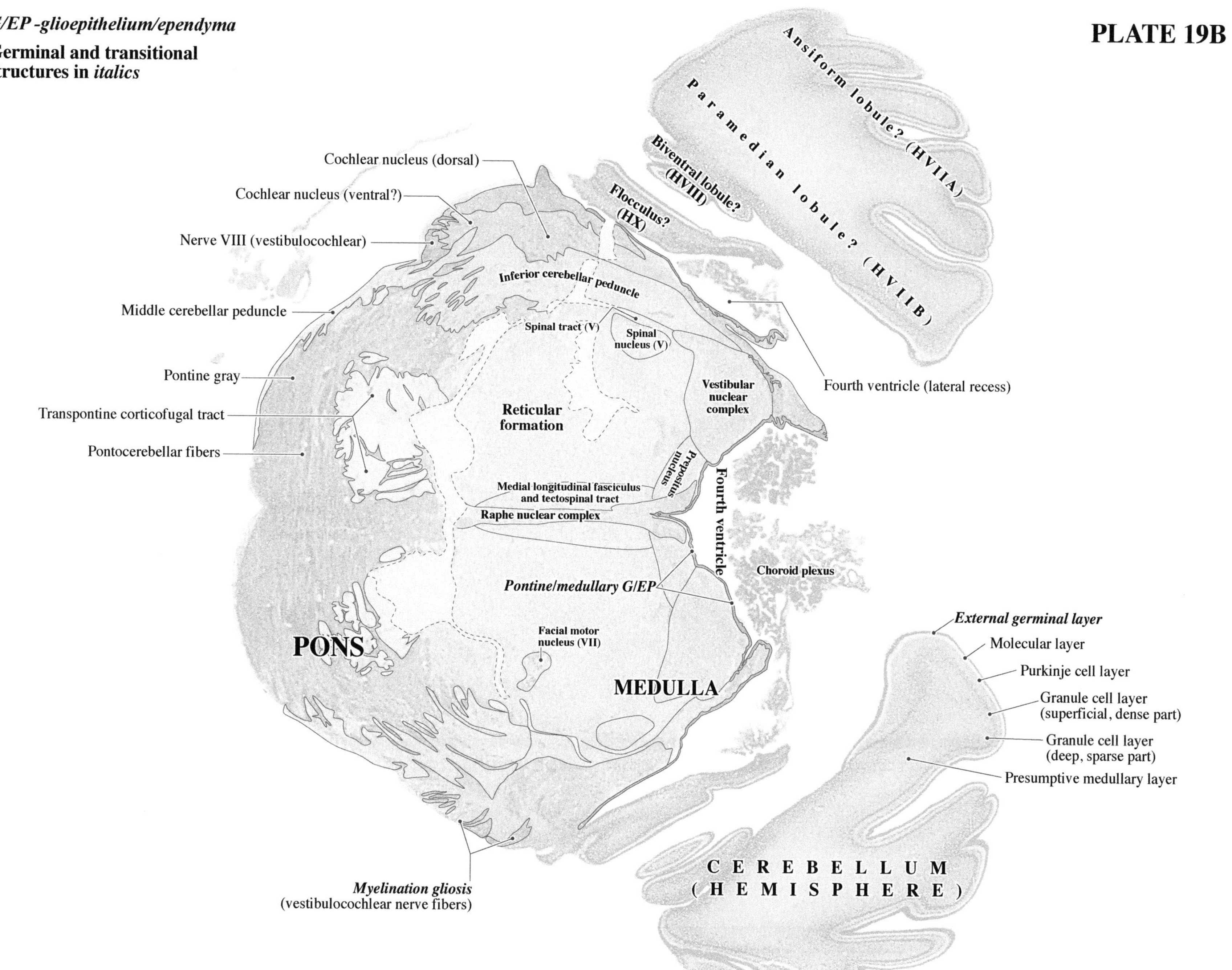

PLATE 20A
CR 190 mm
GW 23, Y197-65
Horizontal
Section 781

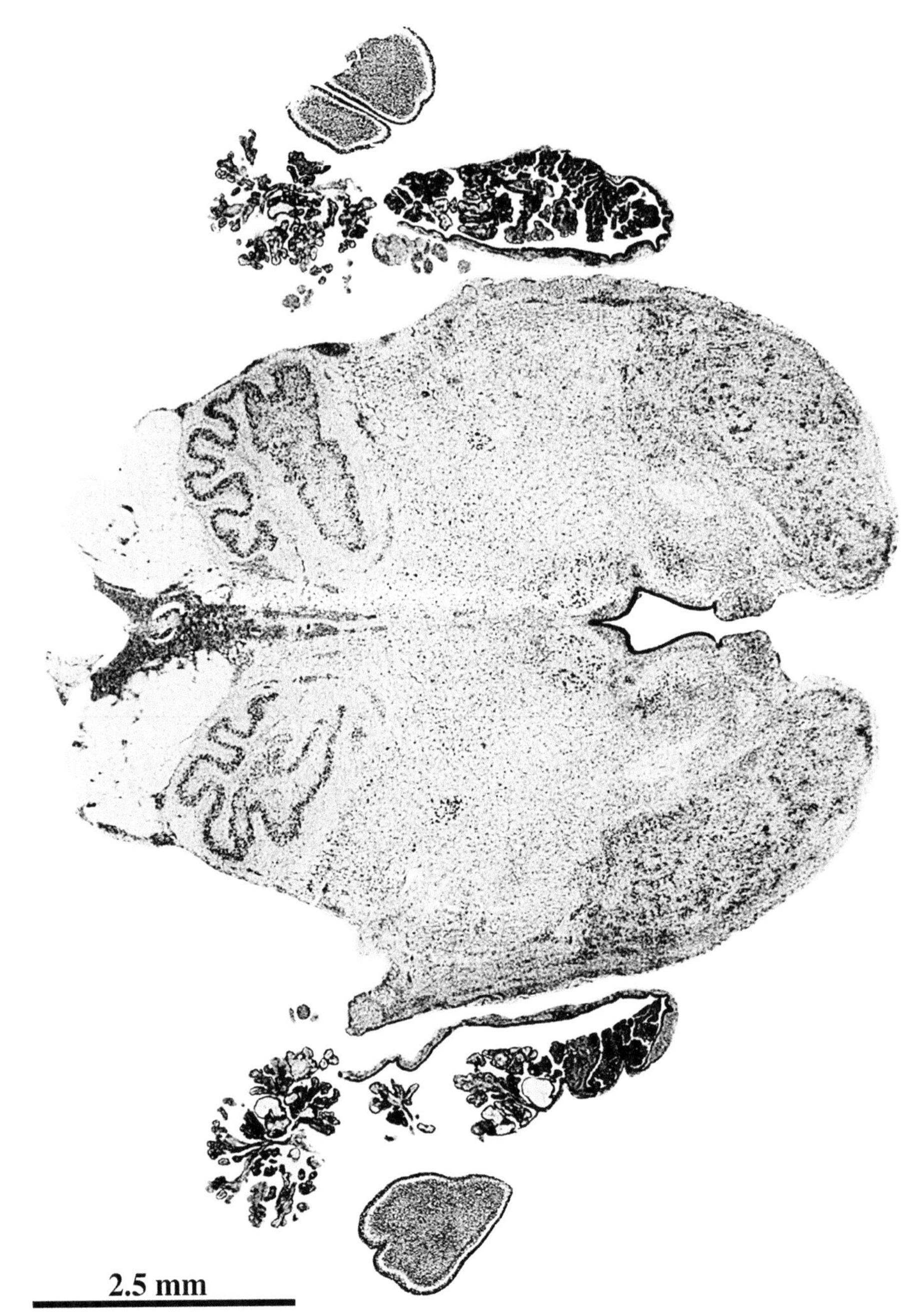

PLATE 20B

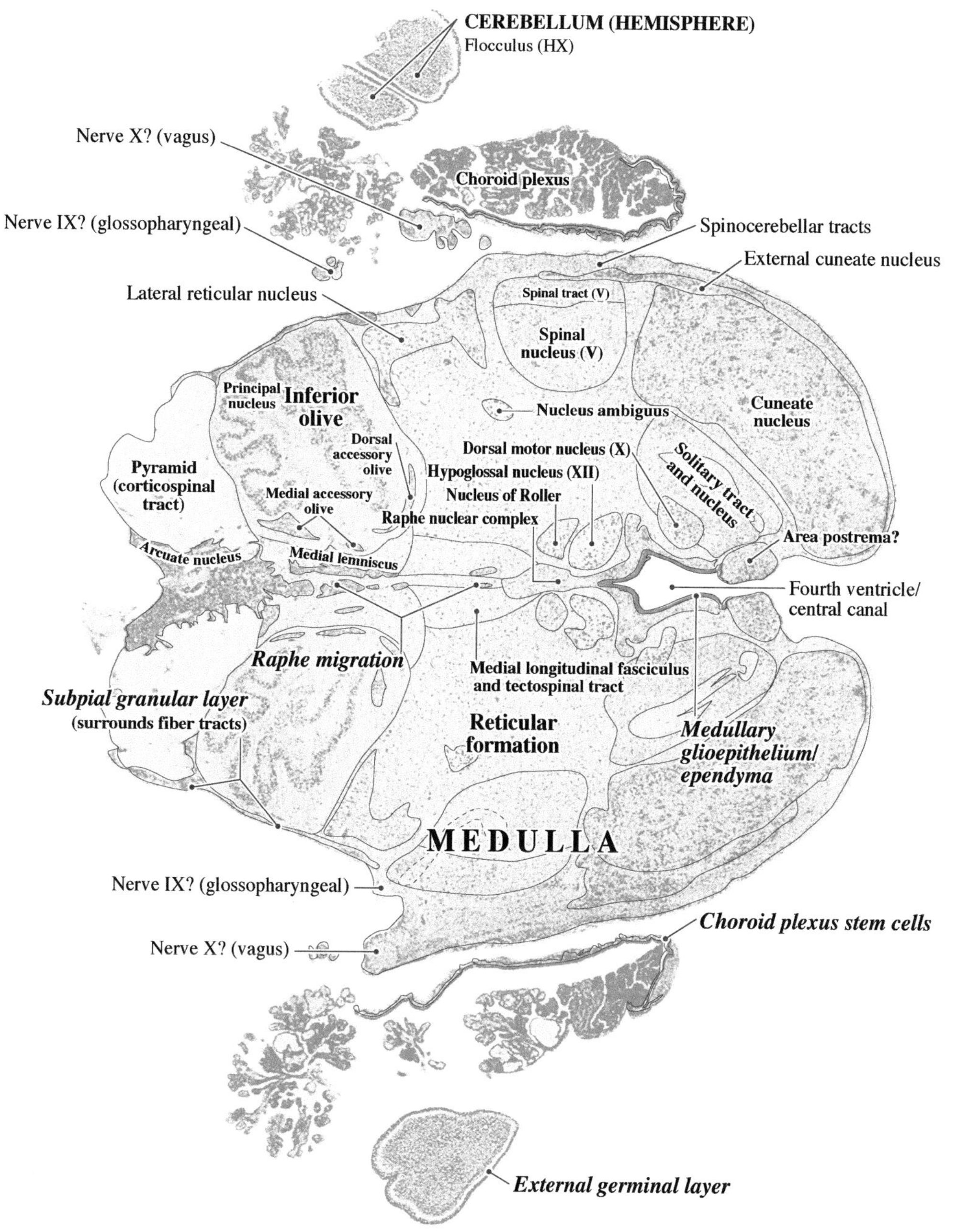

CEREBELLUM (HEMISPHERE)
Flocculus (HX)
Nerve X? (vagus)
Choroid plexus
Nerve IX? (glossopharyngeal)
Spinocerebellar tracts
External cuneate nucleus
Lateral reticular nucleus
Spinal tract (V)
Spinal nucleus (V)
Principal nucleus
Inferior olive
Nucleus ambiguus
Cuneate nucleus
Dorsal accessory olive
Dorsal motor nucleus (X)
Hypoglossal nucleus (XII)
Solitary tract and nucleus
Pyramid (corticospinal tract)
Medial accessory olive
Nucleus of Roller
Raphe nuclear complex
Area postrema?
Arcuate nucleus
Medial lemniscus
Fourth ventricle/ central canal
Raphe migration
Medial longitudinal fasciculus and tectospinal tract
Subpial granular layer
(surrounds fiber tracts)
Reticular formation
Medullary glioepithelium/ ependyma
MEDULLA
Nerve IX? (glossopharyngeal)
Choroid plexus stem cells
Nerve X? (vagus)
External germinal layer

PLATE 21A
CR 190 mm
GW 23, Y197-65
Horizontal
Section 801

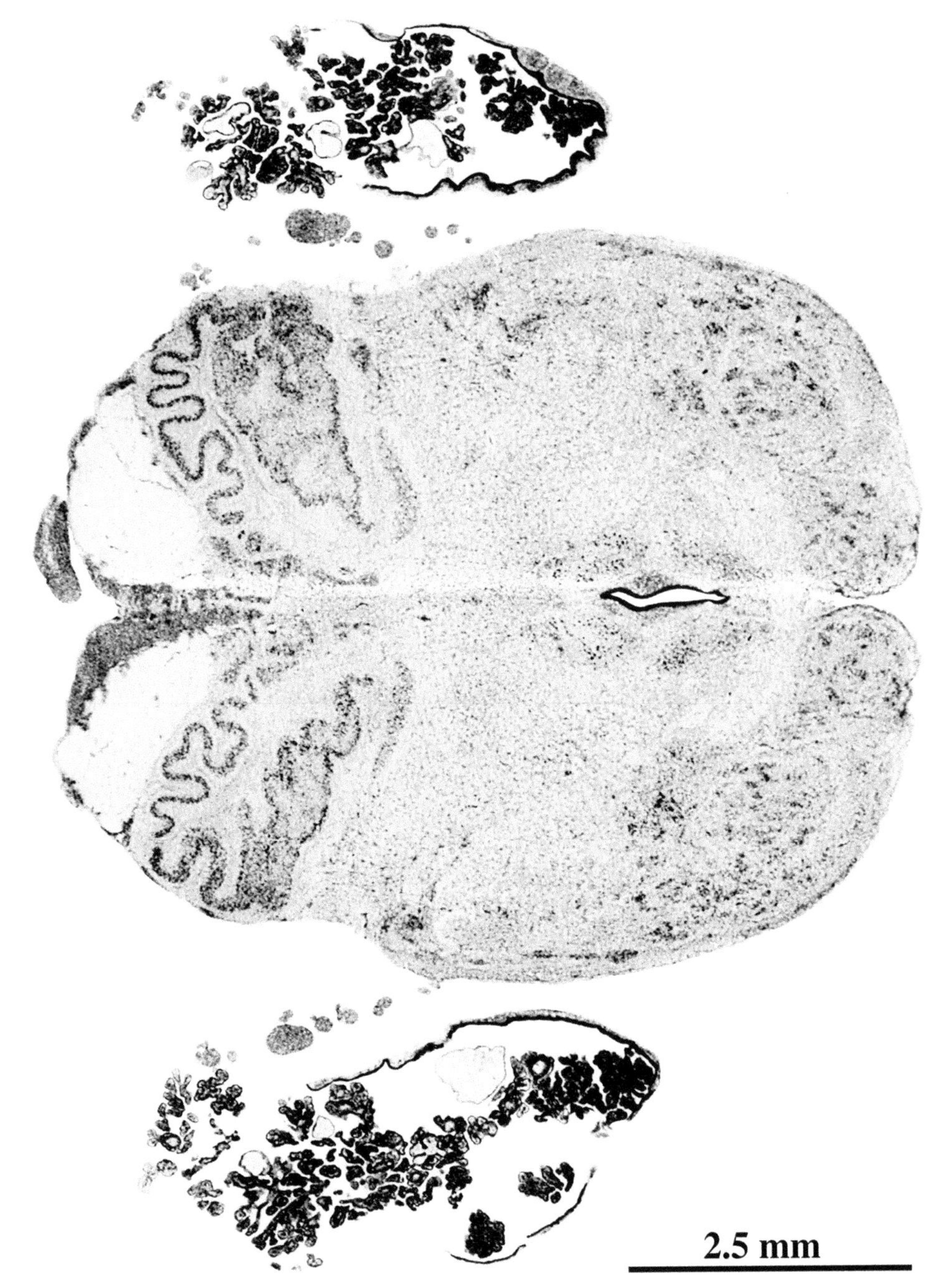

PLATE 21B

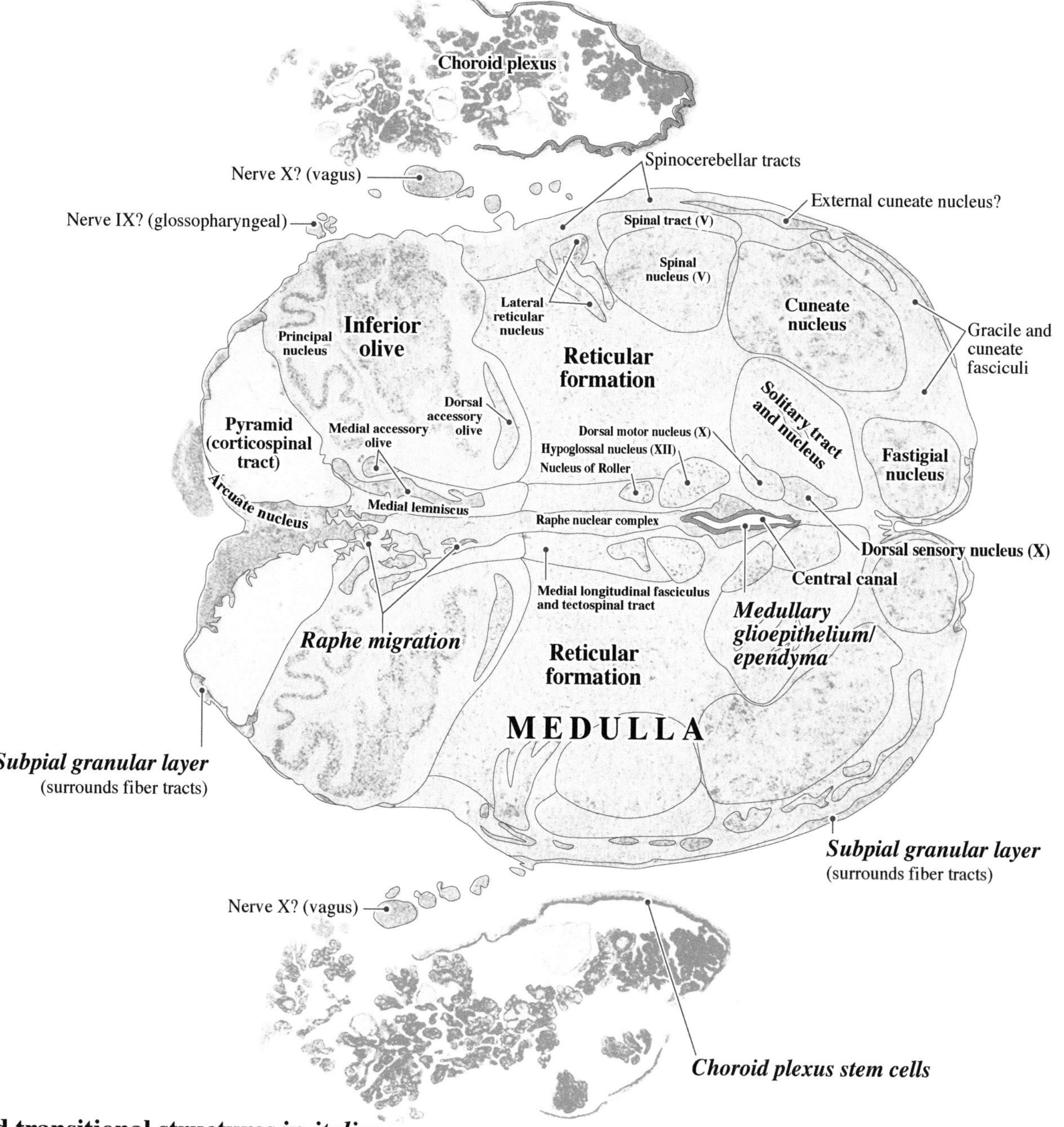

Germinal and transitional structures in *italics*

PLATE 22A
CR 190 mm
GW 23, Y197-65
Horizontal
Section 841

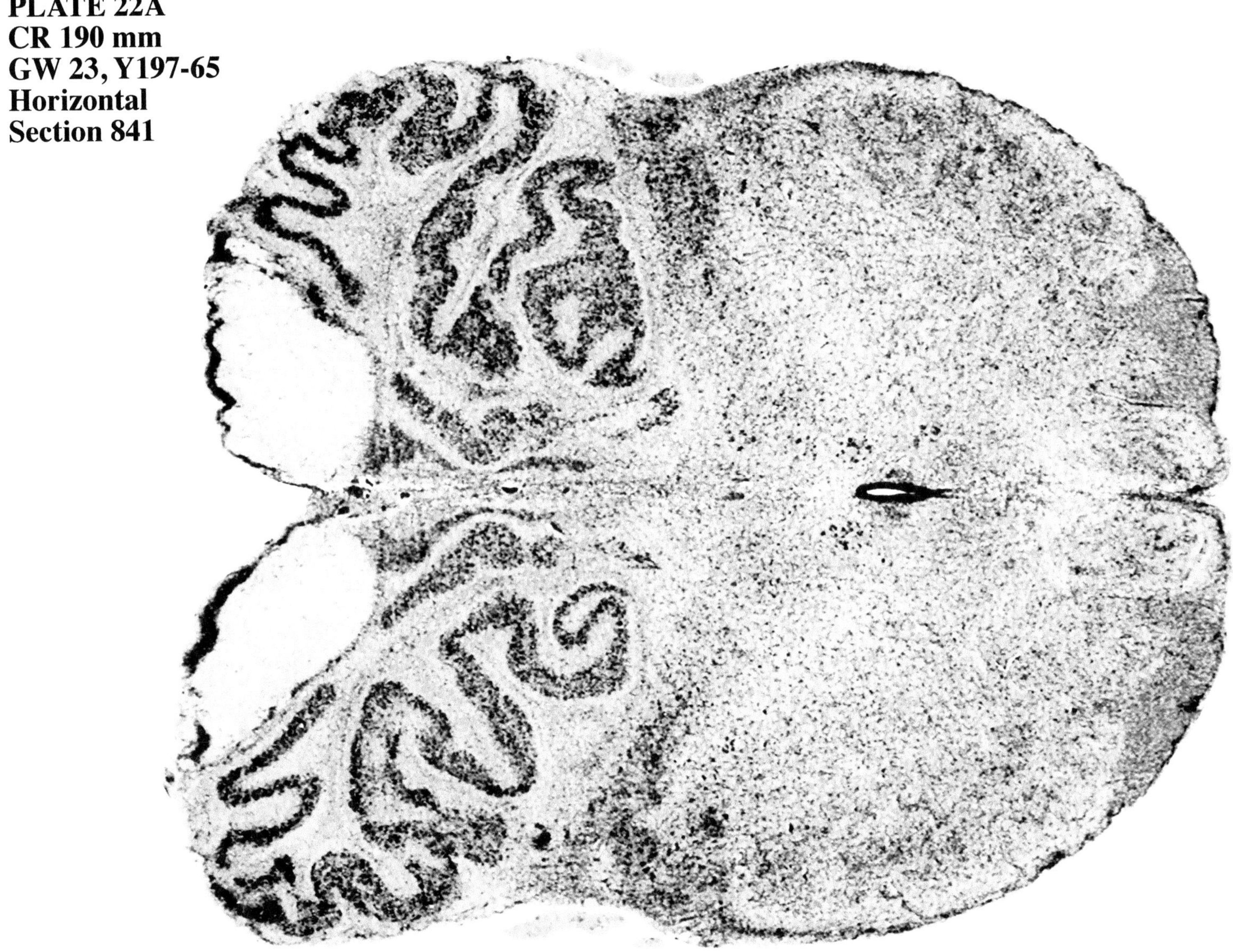

2.5 mm

PLATE 22B

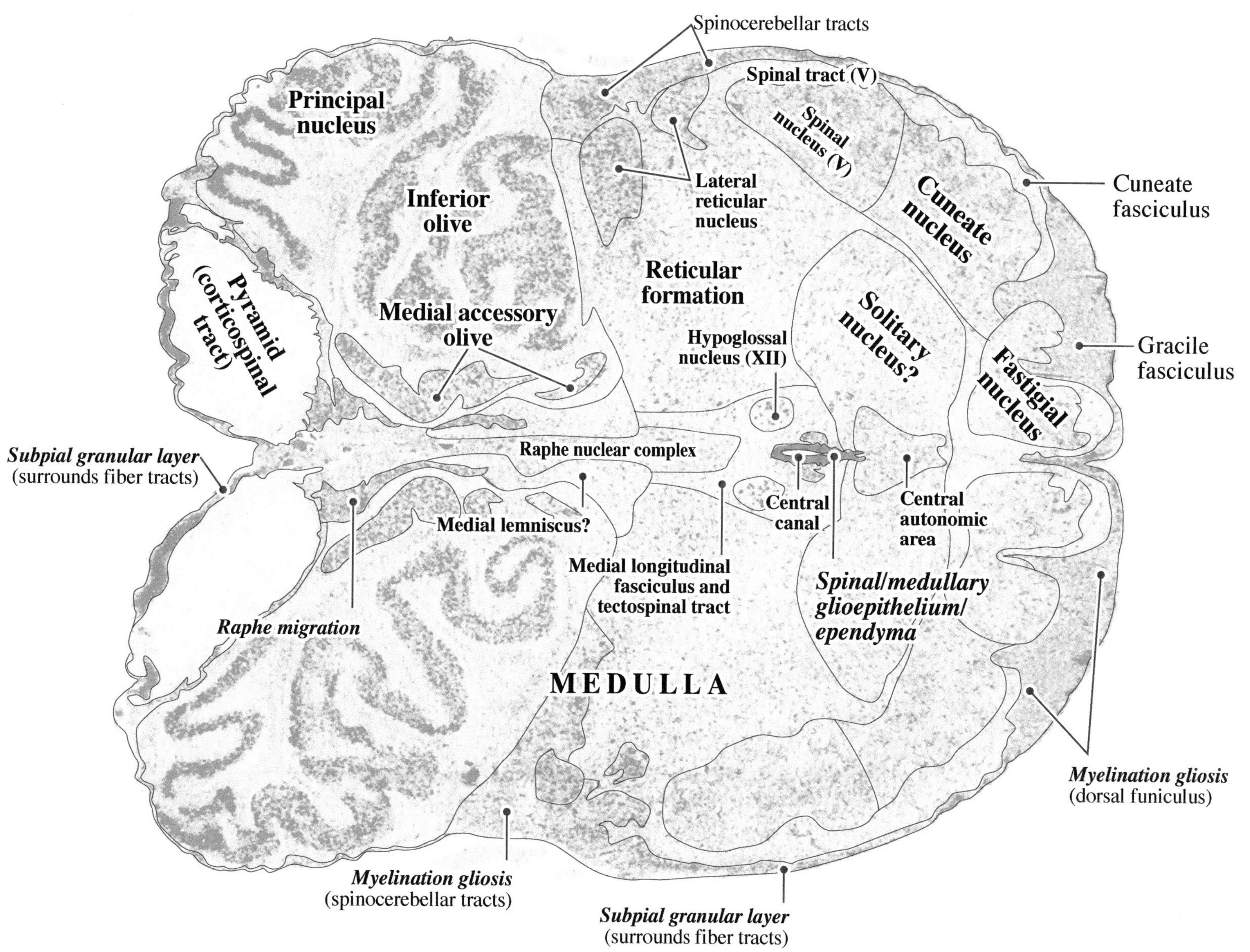

Germinal and transitional structures in *italics*

PLATE 23A
CR 190 mm
GW 20, Y197-65
Horizontal
Section 901

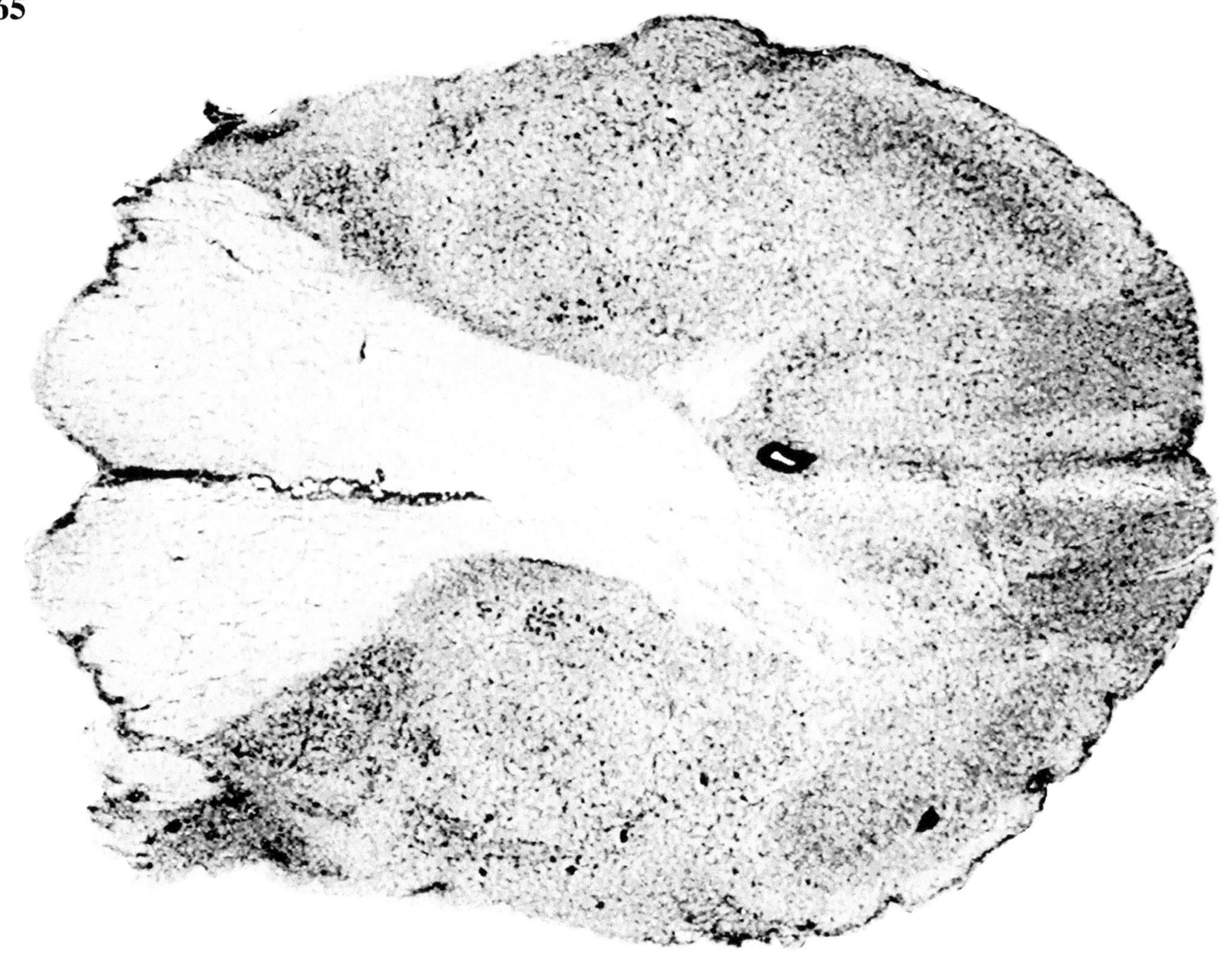

2.5 mm

PLATE 23B

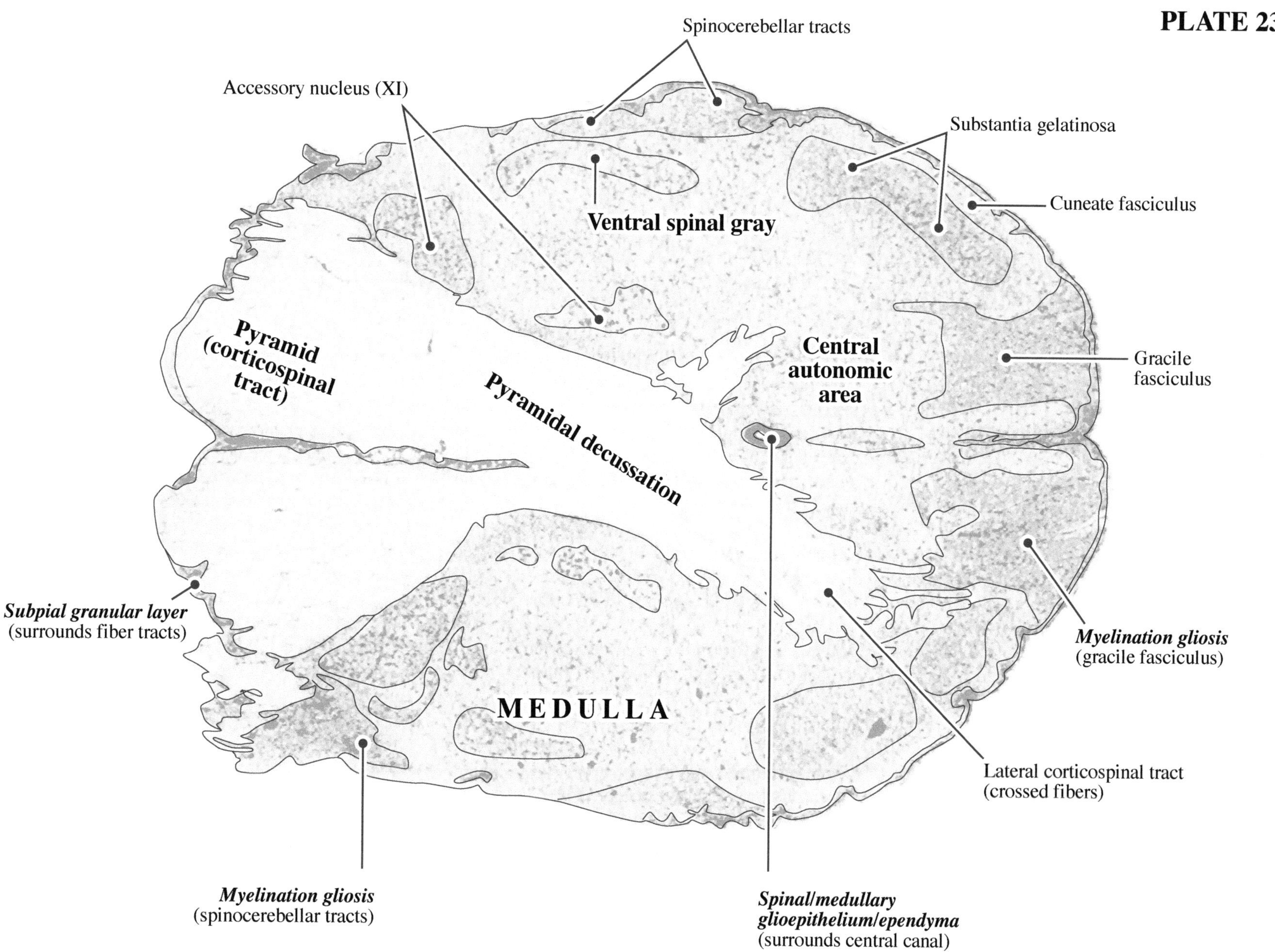

Germinal and transitional structures in *italics*

PART III: Y24-60
CR 205 mm (GW 23.5)
Sagittal

This specimen is a stillborn female fetus (Yakovlev case number RPSL B-24-60; designated here as Y24-60) with a crown-rump length (CR) of 205 mm estimated to be at gestational week (GW) 23.5. The brain was cut in the sagittal plane in 842 sections, 35-μm thick and is classified as a Normative Control in the Yakovlev Collection (Haleem, 1990). Since there is no photograph of this brain before it was embedded and cut, we used the comprehensive atlas that Retzius published in 1896 showing whole fetal brains in medial, lateral, superior, and inferior views and midline sagittally cut brains. **Figure 4** (Retzius, 1896), shows the midline sagittal surface of a brain from a specimen at approximately the same age as Y24-60.

Photographs of 5 different Nissl-stained sections from the right hemisphere are shown at low magnification in **Plates 24–28**. Very-high-magnification views of different regions of the cerebral cortex are shown in **Plates 29** and **30**. The cutting plane of this brain is not parallel to the midline, and many midline sections are too damaged to use. The cerebral cortex was twisted during processing so that the occipital lobe from the left hemisphere is present in the first two levels. The cerebellum has also been displaced so that the section nearest its midline is at the lateral edge of the brainstem (Section 561, **Plate 26**). Because midline sections were damaged, the embryonic structures surrounding the midline ventricular system are not shown in this specimen. Also, no very-high-magnification plates are illustrated for different parts of the vermal cerebellar cortex. In spite of the problems with cutting angles, the brain itself is well-fixed and has excellent histological preservation.

In the cortical regions of the telencephalon, the ***neuroepithelium/subventricular zone*** is generating neocortical neurons destined to reside in superficial cortical layers (mainly II and III). There are regional differences in the thickness of this germinal zone in the parietal, temporal, and occipital lobes, indicating different dynamics of cell proliferation. Many migrating and sojourning neurons and/or glia are visible in all lobes of the cerebral cortex as ***stratified transitional fields (STF)***, with pronounced regional heterogeneity between granular (future sensory) and agranular (future motor) areas. The high-magnification cortical slices show sharp delineation of the various ***STF*** layers. Within the lateral parts of the cerebral cortex, neurons and glia are numerous in the ***lateral migratory stream***. That stream percolates through the claustrum, endopiriform nucleus, external capsule, and uncinate fasciculus, and the cells appear to be heading toward the insular cortex, primary olfactory cortex, temporal cortex, and basolateral parts of the amygdaloid complex. Stem cells of dentate granule cells populate the periphery of the dentate hilus just beneath the granular layer to form the ***subgranular zone*** where dentate granule cells are generated. There is a prominent ***neuroepithelium/subventricular zone*** overlying the nucleus accumbens and striatum where neurons (and glia) are being generated. The striatal portion can be subdivided into anterolateral, anteromedial, and posterior parts. Other structures in the telencephalon, such as the septum and fornix, have only a thin, darkly staining cell layer at the ventricle, and these are presumed to be generating glia, cells of the choroid plexus, and the ependymal lining of the lateral ventricle.

The cerebellum is quite immature. The dentate nucleus is only partially laminated. The entire surface of the cerebellar cortex is covered by the prominent ***external germinal layer*** (***egl***) that is actively producing basket, stellate, and granule cells. Most cortical neurons, including Purkinje cells, are migrating. Lobulation has begun in both the vermis and hemispheres. The ***germinal trigone*** is large at the base of the nodulus and along the floccular peduncle; choroid plexus cells and glia are originating here. The germinal trigone is probably contributing cells to the ***egl***.

GW 23.5 MIDLINE SAGITTAL VIEW

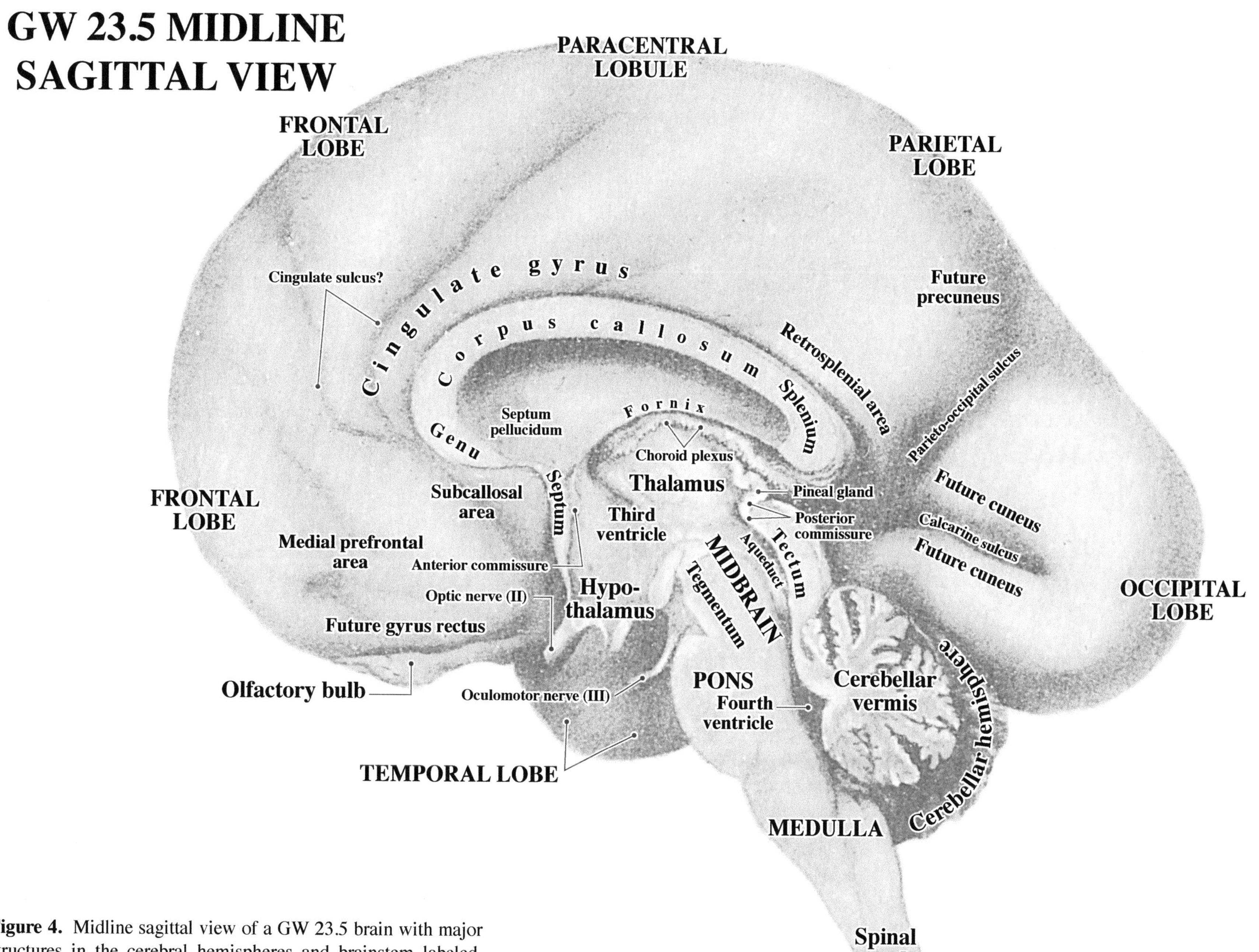

Figure 4. Midline sagittal view of a GW 23.5 brain with major structures in the cerebral hemispheres and brainstem labeled. (This is Figure 5 in Table 11, Volume 2, Retzius, 1896.)

PLATE 24A
CR 205 mm
GW 23.5, Y24-60
Sagittal
Section 461

LAYERS OF THE CORTICAL
STRATIFIED TRANSITIONAL
FIELD (STF)

STF1—Superficial fibrous layer with an early developmental stage ***(t1)*** when many cells are migrating through it, followed by a late stage ***(t2)*** with sparse cells. Endures as the subcortical white matter.

STF2—Upper cellular layer, the last sojourn zone before cells translocate to the cortical plate.

STF3—Honeycomb trilaminar matrix ***(3a, 3b, 3c)*** of cells and fibers found only in granular cortices.

STF4—Complex middle layer with three developmental stages:
t1– fibrous layer without interspersed cells;
t2– cells and fibers intermingle to form striations; ***t3***– fibers endure in the deep white matter.

STF5—Deep cellular layer, the first sojourn zone to appear outside the germinal matrix.

STF6—Late-forming deep layer of callosal fibers outside the germinal matrix.

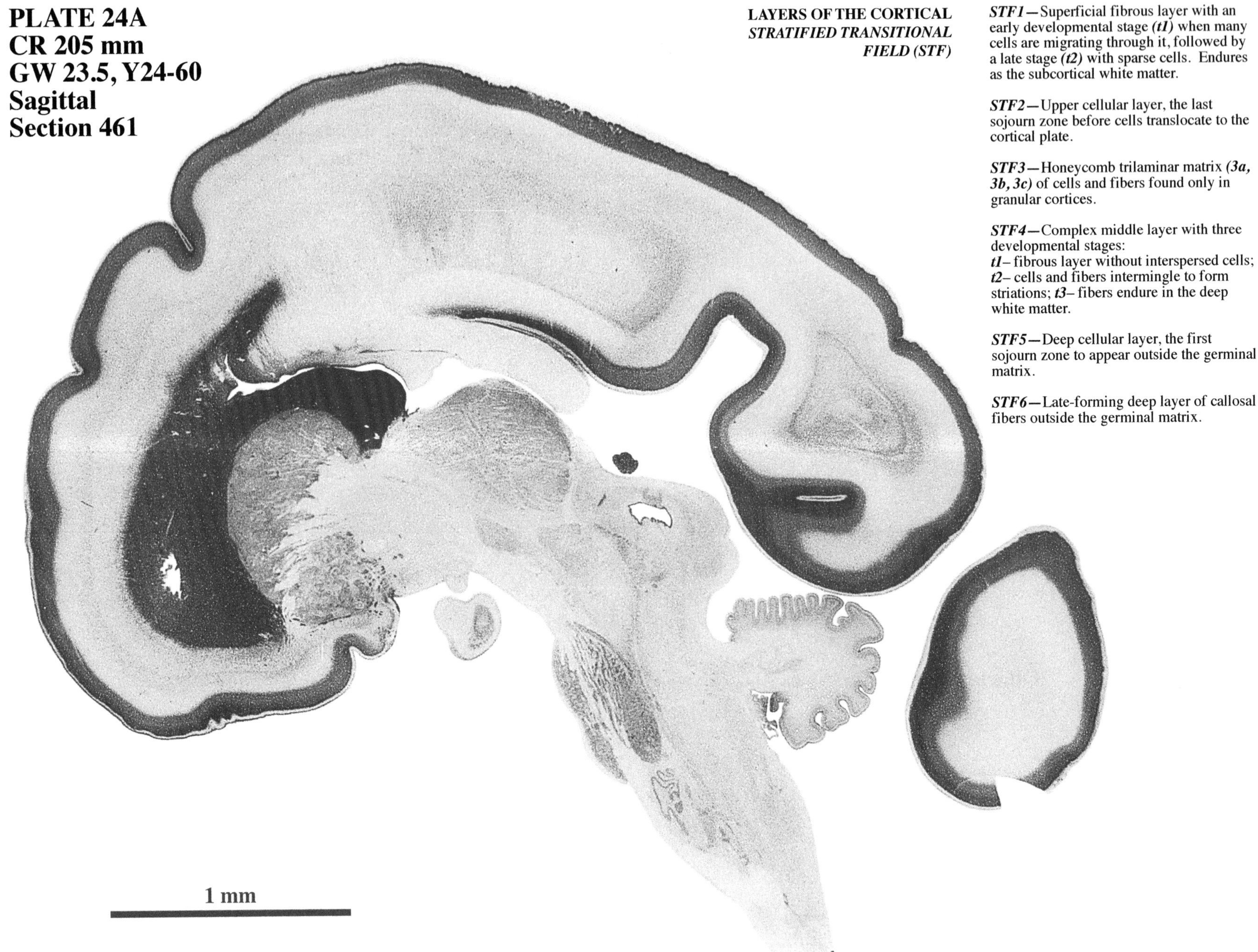

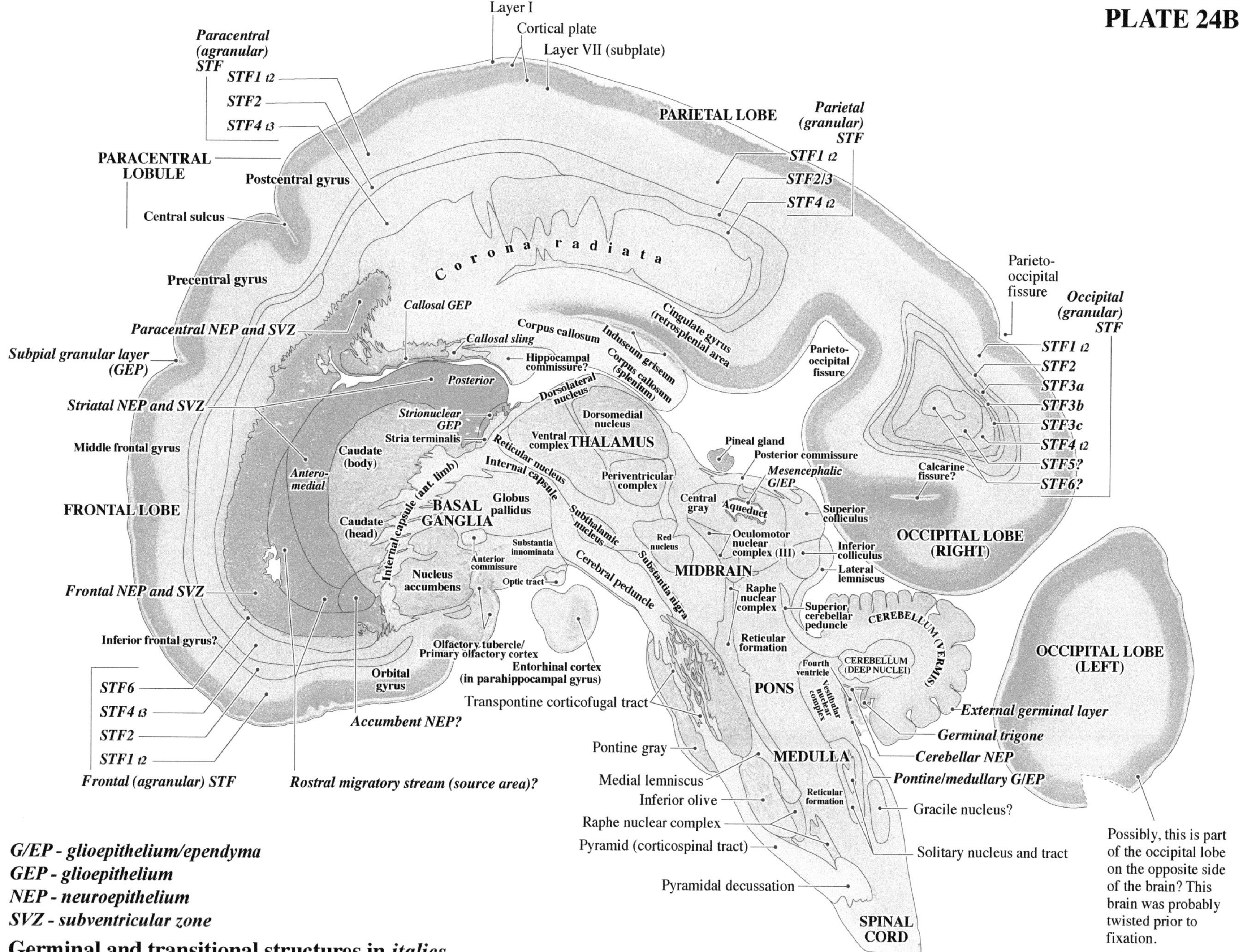
PLATE 24B
Layer I
Cortical plate
Layer VII (subplate)
Paracentral (agranular) STF
STF1 t2
STF2
STF4 t3
PARACENTRAL LOBULE
Postcentral gyrus
Central sulcus
Precentral gyrus
PARIETAL LOBE
Parietal (granular) STF
STF1 t2
STF2/3
STF4 t2
Corona radiata
Callosal GEP
Paracentral NEP and SVZ
Subpial granular layer (GEP)
Callosal sling
Corpus callosum
Hippocampal commissure?
Induseum griseum
Corpus callosum (splenium)
Cingulate gyrus (retrosplenial area)
Parieto-occipital fissure
Parieto-occipital fissure
Occipital (granular) STF
STF1 t2
STF2
STF3a
STF3b
STF3c
STF4 t2
STF5?
STF6?
Posterior
Striatal NEP and SVZ
Strionuclear GEP
Stria terminalis
Dorsolateral nucleus
Dorsomedial nucleus
Ventral complex
THALAMUS
Middle frontal gyrus
Antero-medial
Caudate (body)
Reticular nucleus
Internal capsule
Periventricular complex
Pineal gland
Posterior commissure
Mesencephalic G/EP
Calcarine fissure?
FRONTAL LOBE
Internal capsule (ant. limb)
BASAL GANGLIA
Globus pallidus
Caudate (head)
Central gray
Aqueduct
Superior colliculus
OCCIPITAL LOBE (RIGHT)
Subthalamic nucleus
Red nucleus
Oculomotor nuclear complex (III)
Substantia innominata
Anterior commissure
Inferior colliculus
Nucleus accumbens
Optic tract
Cerebral peduncle
Substantia nigra
MIDBRAIN
Lateral lemniscus
Frontal NEP and SVZ
Raphe nuclear complex
Superior cerebellar peduncle
CEREBELLUM (VERMIS)
Reticular formation
Inferior frontal gyrus?
Olfactory tubercle/ Primary olfactory cortex
Entorhinal cortex (in parahippocampal gyrus)
Fourth ventricle
CEREBELLUM (DEEP NUCLEI)
OCCIPITAL LOBE (LEFT)
Orbital gyrus
PONS
Vestibular nuclear complex
STF6
STF4 t3
STF2
STF1 t2
Transpontine corticofugal tract
External germinal layer
Accumbent NEP?
Germinal trigone
Pontine gray
MEDULLA
Cerebellar NEP
Frontal (agranular) STF
Rostral migratory stream (source area)?
Medial lemniscus
Pontine/medullary G/EP
Inferior olive
Reticular formation
Gracile nucleus?
Raphe nuclear complex
Pyramid (corticospinal tract)
Solitary nucleus and tract
Possibly, this is part of the occipital lobe on the opposite side of the brain? This brain was probably twisted prior to fixation.
Pyramidal decussation
SPINAL CORD
G/EP - glioepithelium/ependyma
GEP - glioepithelium
NEP - neuroepithelium
SVZ - subventricular zone
Germinal and transitional structures in italics

PLATE 25A
CR 205 mm
GW 23.5, Y24-60
Sagittal
Section 501

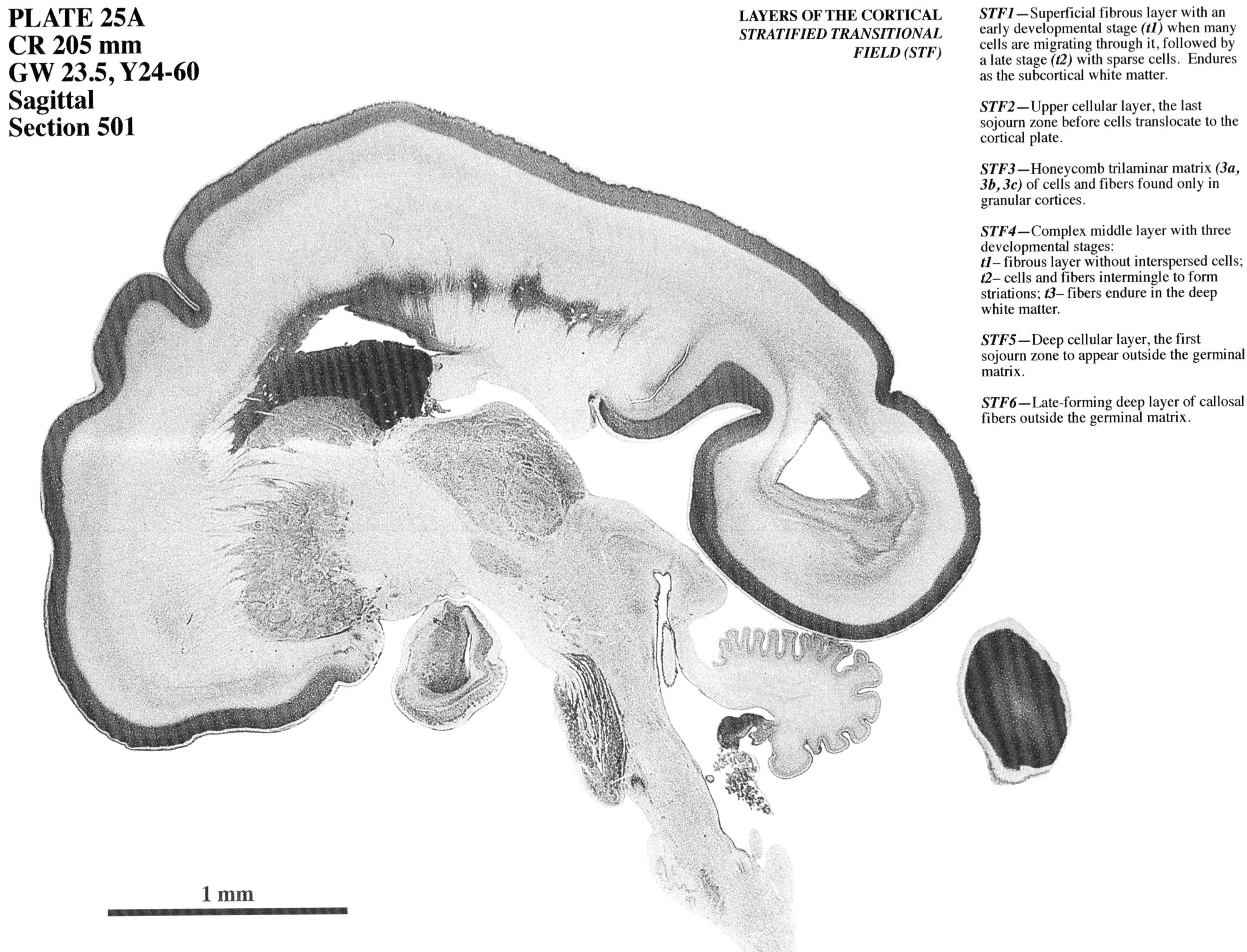

LAYERS OF THE CORTICAL *STRATIFIED TRANSITIONAL FIELD (STF)*

STF1—Superficial fibrous layer with an early developmental stage ***(t1)*** when many cells are migrating through it, followed by a late stage ***(t2)*** with sparse cells. Endures as the subcortical white matter.

STF2—Upper cellular layer, the last sojourn zone before cells translocate to the cortical plate.

STF3—Honeycomb trilaminar matrix ***(3a, 3b, 3c)*** of cells and fibers found only in granular cortices.

STF4—Complex middle layer with three developmental stages:
t1– fibrous layer without interspersed cells; ***t2***– cells and fibers intermingle to form striations; ***t3***– fibers endure in the deep white matter.

STF5—Deep cellular layer, the first sojourn zone to appear outside the germinal matrix.

STF6—Late-forming deep layer of callosal fibers outside the germinal matrix.

PLATE 25B

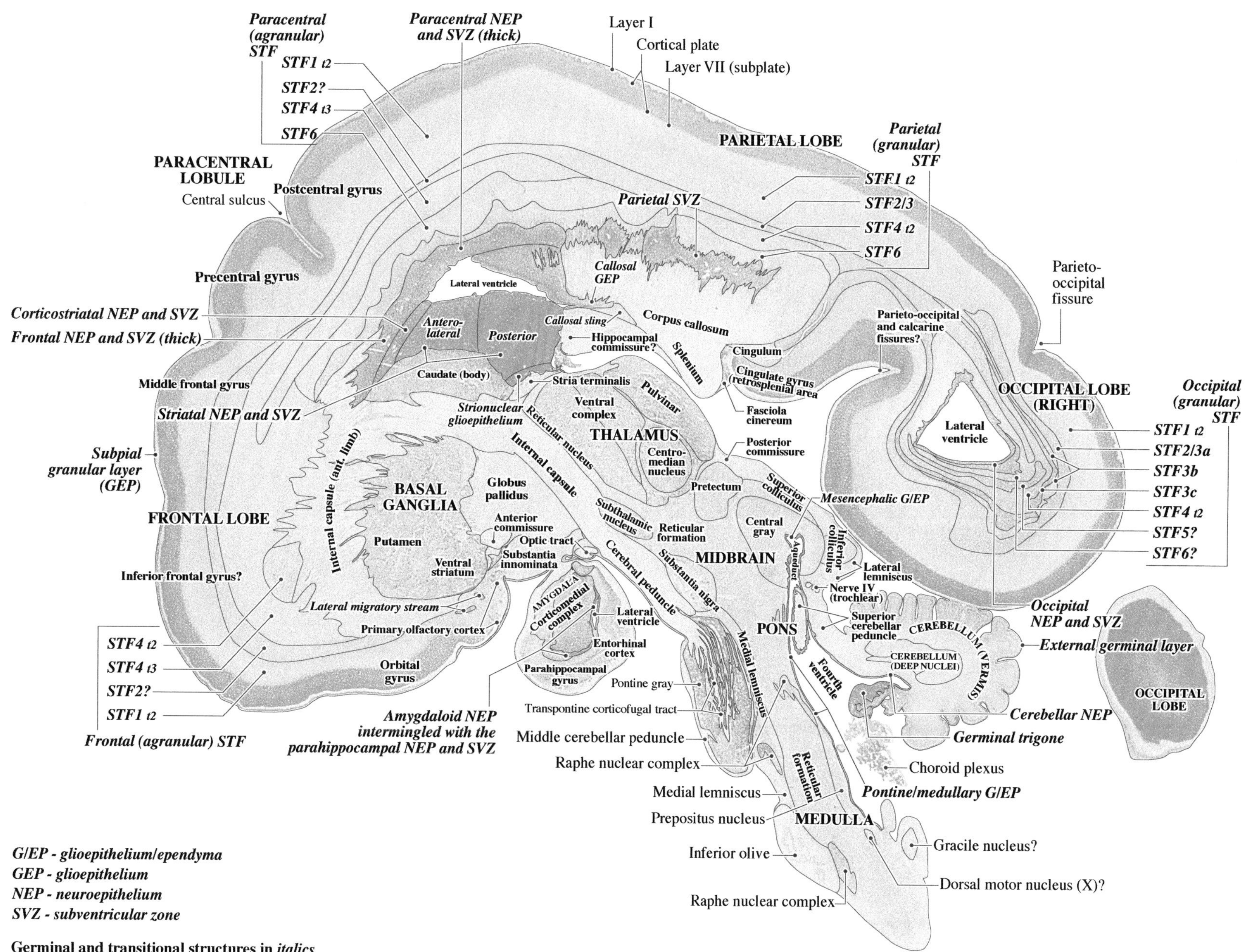

G/EP - glioepithelium/ependyma
GEP - glioepithelium
NEP - neuroepithelium
SVZ - subventricular zone

Germinal and transitional structures in *italics*

PLATE 26A
CR 205 mm
GW 23.5, Y24-60
Sagittal
Section 561

LAYERS OF THE CORTICAL
STRATIFIED TRANSITIONAL FIELD (STF)

STF1—Superficial fibrous layer with an early developmental stage ***(t1)*** when many cells are migrating through it, followed by a late stage ***(t2)*** with sparse cells. Endures as the subcortical white matter.

STF2—Upper cellular layer, the last sojourn zone before cells translocate to the cortical plate.

STF3—Honeycomb trilaminar matrix ***(3a, 3b, 3c)*** of cells and fibers found only in granular cortices.

STF4—Complex middle layer with three developmental stages:
t1– fibrous layer without interspersed cells;
t2– cells and fibers intermingle to form striations; ***t3***– fibers endure in the deep white matter.

STF5—Deep cellular layer, the first sojourn zone to appear outside the germinal matrix.

STF6—Late-forming deep layer of callosal fibers outside the germinal matrix.

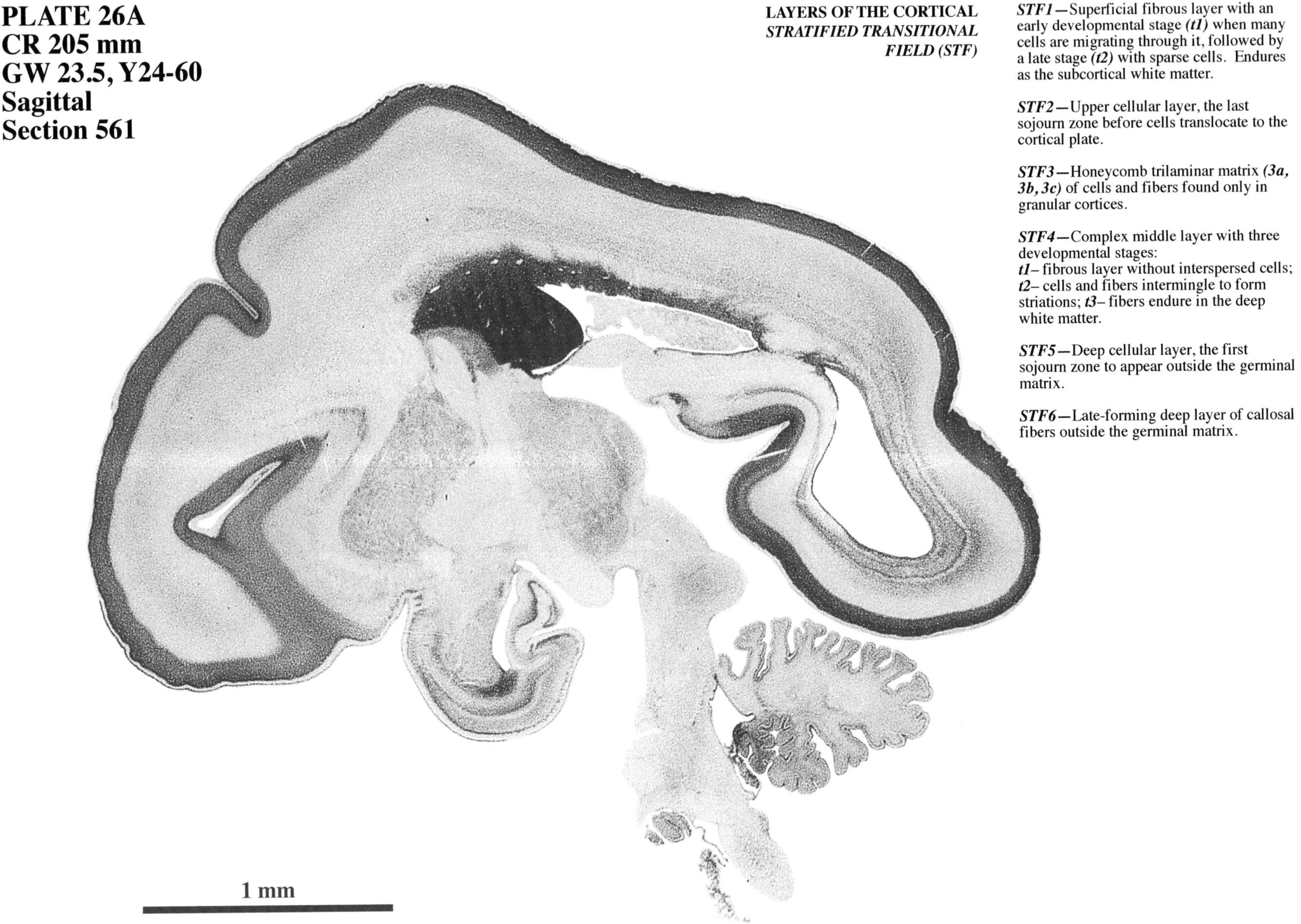

PLATE 26B

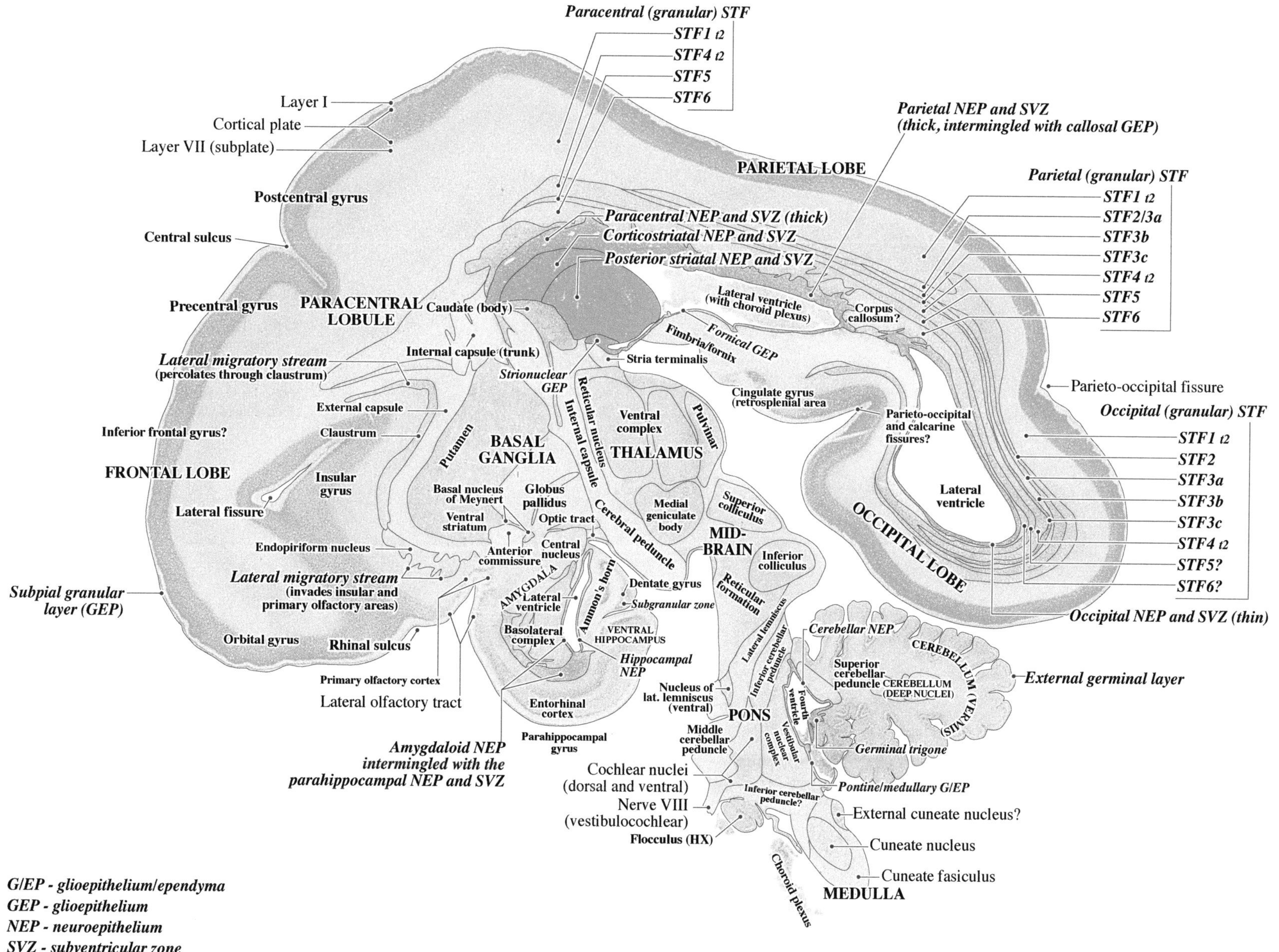

G/EP - glioepithelium/ependyma
GEP - glioepithelium
NEP - neuroepithelium
SVZ - subventricular zone

Germinal and transitional structures in ***italics***

PLATE 27A
CR 205 mm
GW 23.5, Y24-60
Sagittal
Section 621

LAYERS OF THE CORTICAL *STRATIFIED TRANSITIONAL FIELD (STF)*

STF1*—**Superficial fibrous layer with an early developmental stage ***(t1) when many cells are migrating through it, followed by a late stage ***(t2)*** with sparse cells. Endures as the subcortical white matter.

***STF2*—**Upper cellular layer, the last sojourn zone before cells translocate to the cortical plate.

STF3*—**Honeycomb trilaminar matrix ***(3a, 3b, 3c) of cells and fibers found only in granular cortices.

***STF4*—**Complex middle layer with three developmental stages:
t1– fibrous layer without interspersed cells;
t2– cells and fibers intermingle to form striations; ***t3***– fibers endure in the deep white matter.

***STF5*—**Deep cellular layer, the first sojourn zone to appear outside the germinal matrix.

***STF6*—**Late-forming deep layer of callosal fibers outside the germinal matrix.

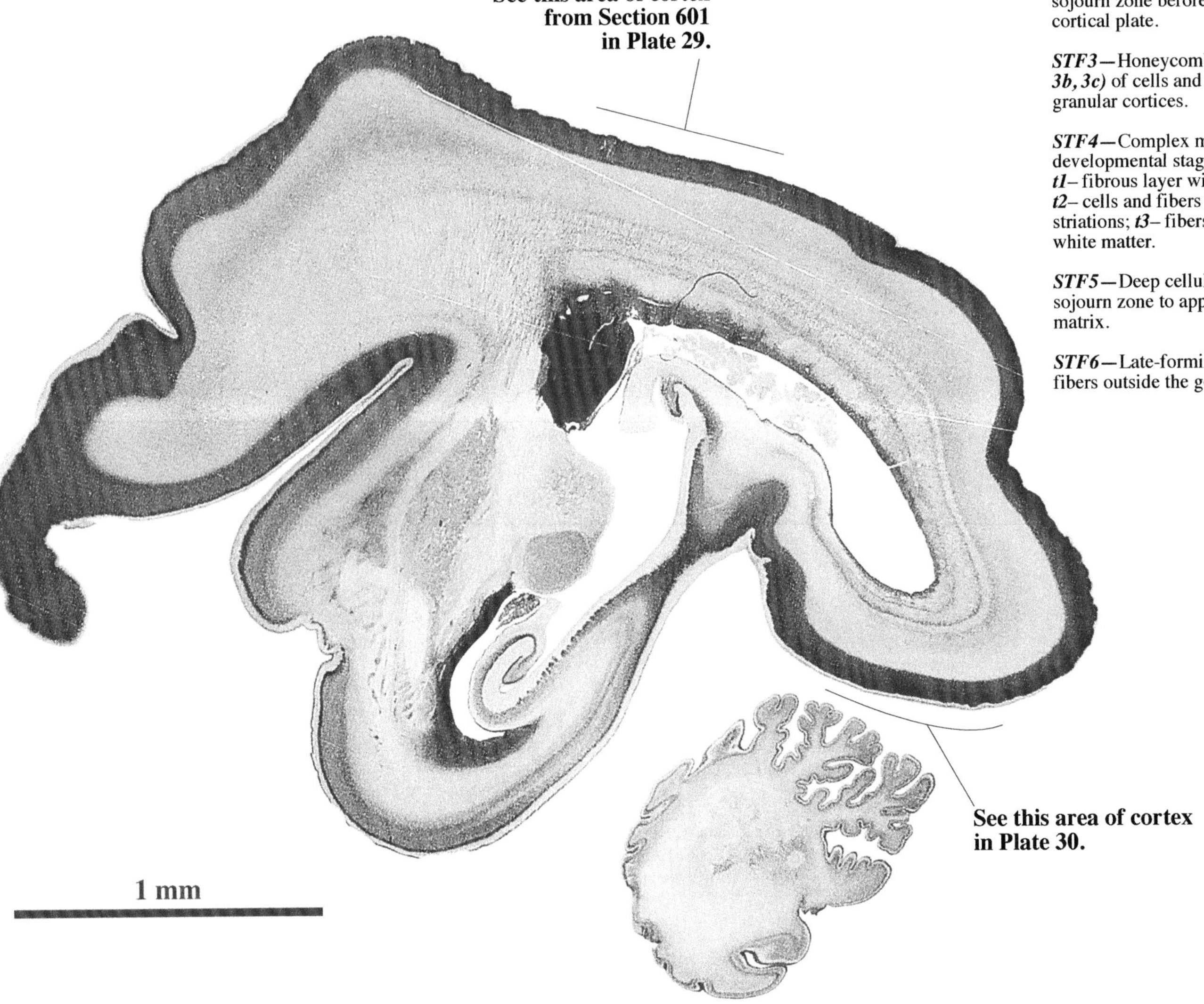

PLATE 27B

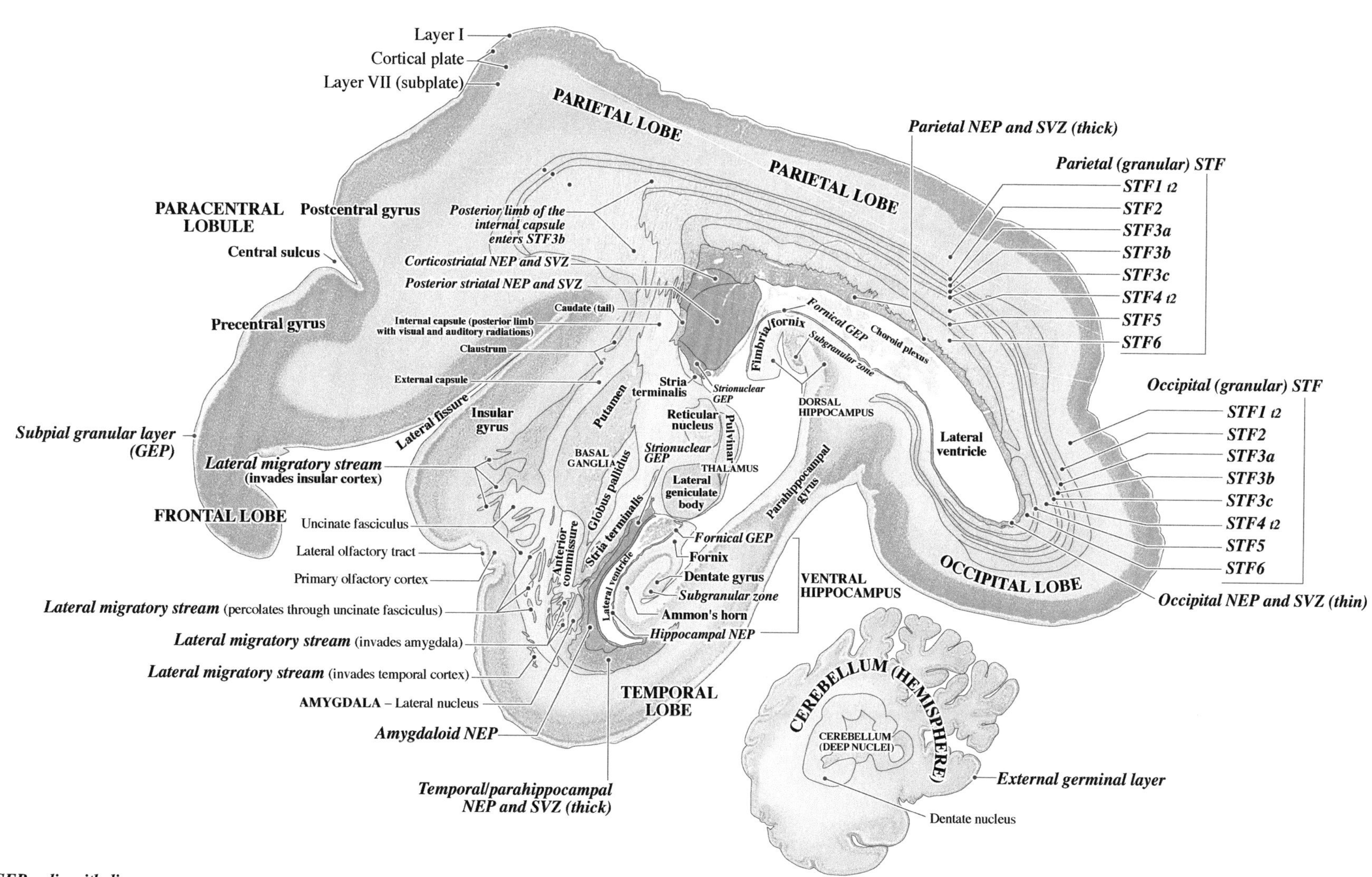

GEP - glioepithelium
NEP - neuroepithelium
SVZ - subventricular zone

Germinal and transitional structures in *italics*

PLATE 28A
CR 205 mm
GW 23.5, Y24-60
Sagittal
Section 681

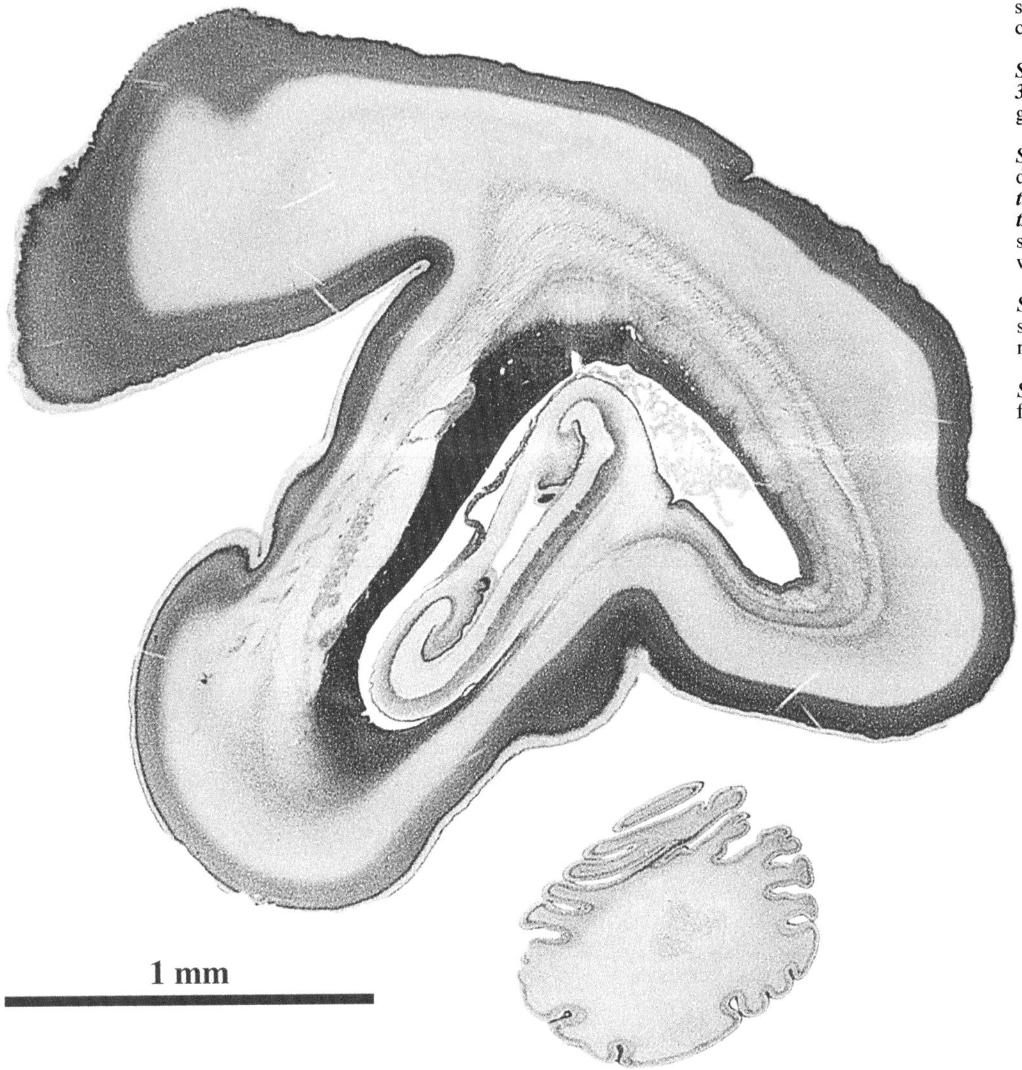

LAYERS OF THE CORTICAL
STRATIFIED TRANSITIONAL
FIELD (STF)

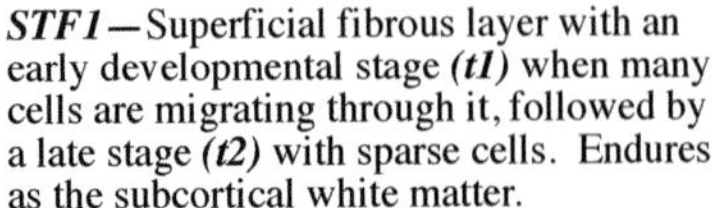

***STF1*—Superficial fibrous layer with an early developmental stage *(t1)* when many cells are migrating through it, followed by a late stage *(t2)* with sparse cells. Endures as the subcortical white matter.

***STF2*—Upper cellular layer, the last sojourn zone before cells translocate to the cortical plate.

STF3*—Honeycomb trilaminar matrix ***(3a, 3b, 3c) of cells and fibers found only in granular cortices.

***STF4*—Complex middle layer with three developmental stages:
t1– fibrous layer without interspersed cells; ***t2***– cells and fibers intermingle to form striations; ***t3***– fibers endure in the deep white matter.

***STF5*—Deep cellular layer, the first sojourn zone to appear outside the germinal matrix.

***STF6*—Late-forming deep layer of callosal fibers outside the germinal matrix.

1 mm

PLATE 28B

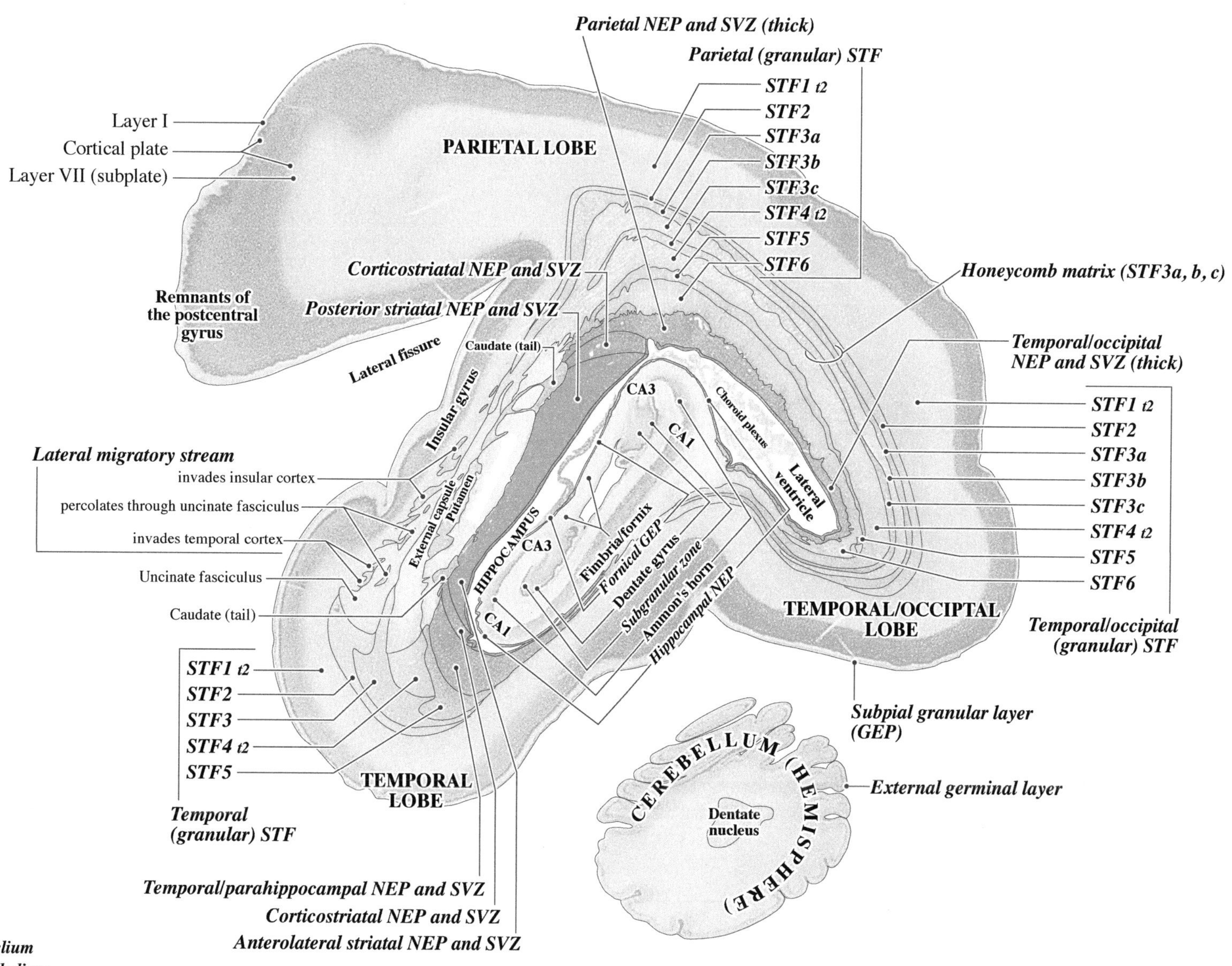

GEP - glioepithelium
NEP - neuroepithelium
SVZ - subventricular zone

Germinal and transitional structures in *italics*

PLATE 29
CR 205 mm
GW 23.5, Y24-60
Sagittal
Section 601
PARIETAL CORTEX

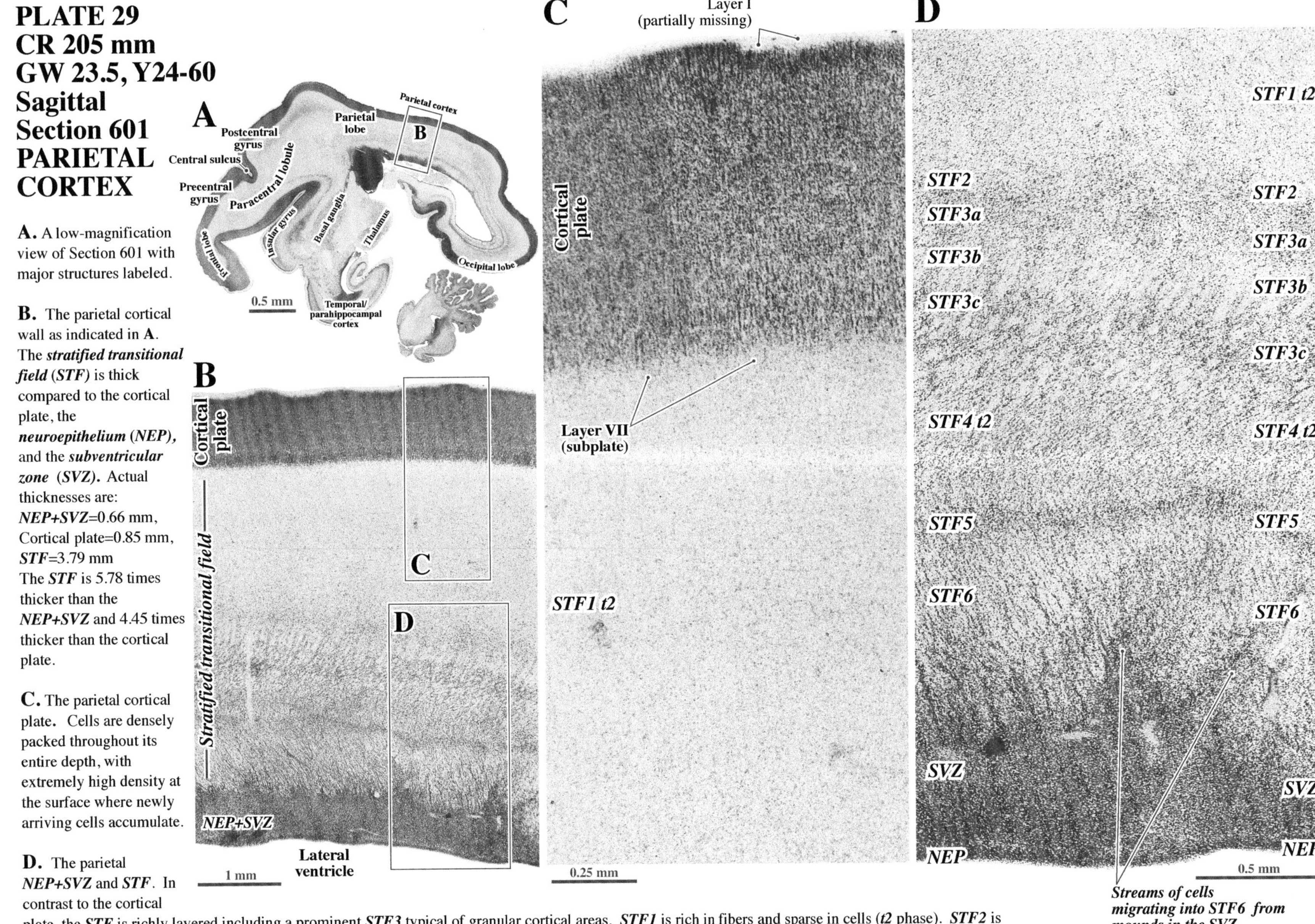

A. A low-magnification view of Section 601 with major structures labeled.

B. The parietal cortical wall as indicated in **A**. The ***stratified transitional field*** (*STF*) is thick compared to the cortical plate, the ***neuroepithelium*** (*NEP*), and the ***subventricular zone*** (*SVZ*). Actual thicknesses are: *NEP+SVZ*=0.66 mm, Cortical plate=0.85 mm, *STF*=3.79 mm The *STF* is 5.78 times thicker than the *NEP+SVZ* and 4.45 times thicker than the cortical plate.

C. The parietal cortical plate. Cells are densely packed throughout its entire depth, with extremely high density at the surface where newly arriving cells accumulate.

D. The parietal *NEP+SVZ* and *STF*. In contrast to the cortical plate, the *STF* is richly layered including a prominent ***STF3*** typical of granular cortical areas. ***STF1*** is rich in fibers and sparse in cells (*t2* phase). ***STF2*** is mainly a transition zone between ***STF1*** and ***STF3***. ***STF3*** is subdivided into cell-rich ***a*** and ***c*** sublayers above and below a middle fibrous layer ***(b)*** that has a ***honeycomb matrix*** of radially arrayed cells between the fibers. ***STF4*** is a broad layer where cells intermingle with fibers in the **t2** developmental stage. Presumably, many of these fibers are afferents from the somatosensory nuclei of the thalamus. ***STF5*** is a narrow, dense cell sojourn zone where new neurons may await the arrival of specific afferents. ***STF6*** is filled with fibers that extend toward the corpus callosum.

PLATE 30
CR 205 mm
GW 23.5
Y24-60
Sagittal
Section 621
OCCIPITAL CORTEX

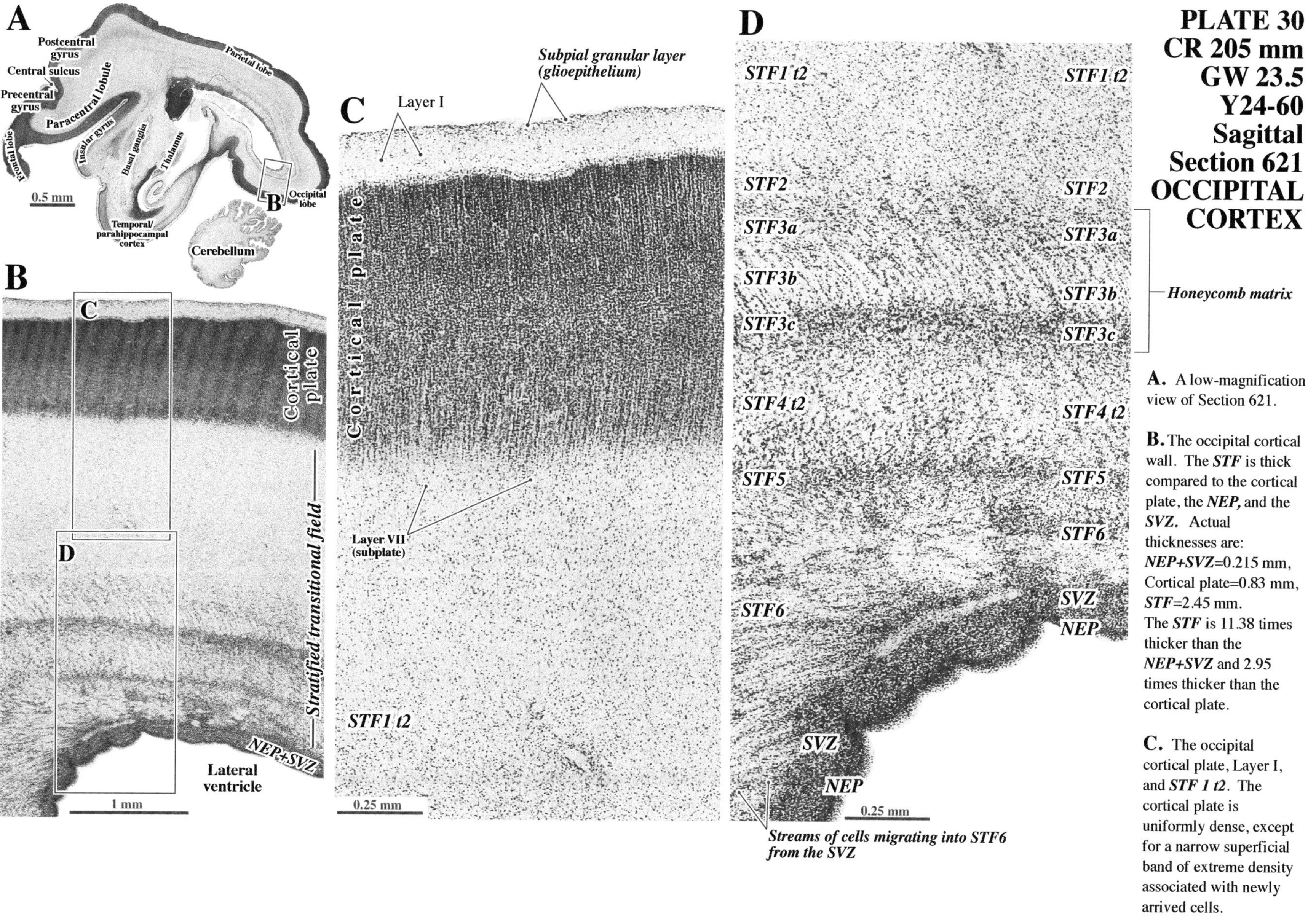

A. A low-magnification view of Section 621.

B. The occipital cortical wall. The ***STF*** is thick compared to the cortical plate, the ***NEP,*** and the ***SVZ.*** Actual thicknesses are: ***NEP+SVZ***=0.215 mm, Cortical plate=0.83 mm, ***STF***=2.45 mm. The ***STF*** is 11.38 times thicker than the ***NEP+SVZ*** and 2.95 times thicker than the cortical plate.

C. The occipital cortical plate, Layer I, and ***STF 1 t2***. The cortical plate is uniformly dense, except for a narrow superficial band of extreme density associated with newly arrived cells.

D. The occipital ***NEP, SVZ,*** and ***STF***. As in Section 601 (**Plate 29**), there are six layers in the ***STF. STF1 t2*** and ***STF2*** are similar to parietal cortex. But the ***a, b,*** and ***c*** sublayers of ***STF3*** are more definite than in parietal cortex. ***STF3*** is consistently most prominent in the occipital cortex where Layer IV becomes most dense in the mature cortex. There are many regularly spaced streams of radially migrating cells traversing ***STF3b*** that constitute the ***honeycomb matrix*** around fibers in the layer.

PART IV: Y94-62 CR 210 mm (GW 24) Frontal

This specimen is a stillborn female fetus (Yakovlev case number RPSL W-94-62; referred to as Y94-62) with a crown-rump length (CR) o f 210 mm estimated to be at gestational week (GW) 24. The brain was cut in the coronal (frontal) plane in 1,452 sections, 35-μm thick and is classified as a Normative Control in the Yakovlev Collection (Haleem, 1990). Since there is no photograph of this brain before it was embedded and cut, we turned to the comprehensive atlas that Retzius published in 1896 showing whole fetal brains in medial, lateral, superior, and inferior views and midline sagittally cut brains. **Figure 5**, (Retzius, 1896), shows the exterior of a brain from a specimen at the same age as Y94-62, along with the approximate cutting angle of the sections. Photographs of 15 entire Nissl-stained sections are illustrated (**Plates 31-45**). **Plates 46-58** show high-magnification views of the central brain. Different areas of the cerebral cortex are shown at still higher magnification in **Plates 59-62**.

Y94-62 contains several immature structures. In the cortical regions of the telencephalon, the ***neuroepithelium/subventricular zone*** is generating neocortical neurons destined to reside in superficial cortical layers (mainly Layers II and III). Many migrating and sojourning neurons and/or glia are visible in ***stratified transitional fields*** in all lobes of the cerebral cortex, with pronounced regional heterogeneity between granular (future sensory) and agranular (future motor) areas. Many neurons, glia, and their mitotic precursor cells are still migrating through the olfactory peduncle toward the olfactory bulb (***rostral migratory stream***) from a presumed source in the germinal matrix at the junction between the cerebral cortex and striatum (***cortical/striatal neuroepithelium and subventricular zone***). Within the lateral parts of the cerebral cortex, streams of neurons and glia are numerous in the ***lateral migratory stream***. That stream percolates through the claustrum, endopiriform nucleus, external capsule, and uncinate fasciculus, and the cells appear to be heading toward the insular cortex, primary olfactory cortex, temporal cortex, and basolateral parts of the amygdaloid complex. In the hippocampus, cells are entering Ammon's horn pyramidal layer via the ***ammonic migratory stream***, and granule cells and their precursors are migrating to the dentate gyrus in the ***dentate migratory stream***. Stem cells populate the periphery of the dentate hilus just beneath the granular layer to form the ***subgranular zone*** where they generate dentate granule cells. There is a prominent ***neuroepithelium/subventricular zone*** overlying the nucleus accumbens and striatum that generates neurons (and glia). The ***striatal neuroepithelium and subventricular zone*** can be subdivided into anterolateral, anteromedial, and posterior parts. The septum has only a thin, dense layer at the ventricle that generates glia and the ependymal lining of the lateral ventricle.

Neurons in most diencephalic structures appear to be settled and are maturing; the third ventricle is lined by a thin ***glioepithelium/ependyma***. Lamination in the lateral geniculate body is just beginning. In the midbrain, pons, and medulla, there is a ***glioepithelium/ependyma*** lining the cerebral aqueduct and fourth ventricle. However, there is evidence of immaturity in the brainstem parenchyma. For example, the transpontine corticofugal tract is more discrete in the pontine gray, and the pontine gray itself is larger than younger specimens. There is an extensive ***raphe migration*** in the midline of the posterior pons/anterior medulla blending with the medullary arcuate nucleus on the ventromedial surface of the pyramids.

The cerebellar dentate nucleus is only partially laminated. The entire surface of the cerebellar cortex is covered by the prominent ***external germinal layer*** that is actively producing basket, stellate, and granule cells. Lamination in the cortex is primordial; some Purkinje cells and most granule cells are migrating. Lobulation has begun in both the vermis and hemispheres continues to progress. The ***germinal trigone*** is large at the base of the nodulus and along the floccular peduncle; choroid plexus stem cells and glia are probably originating there.

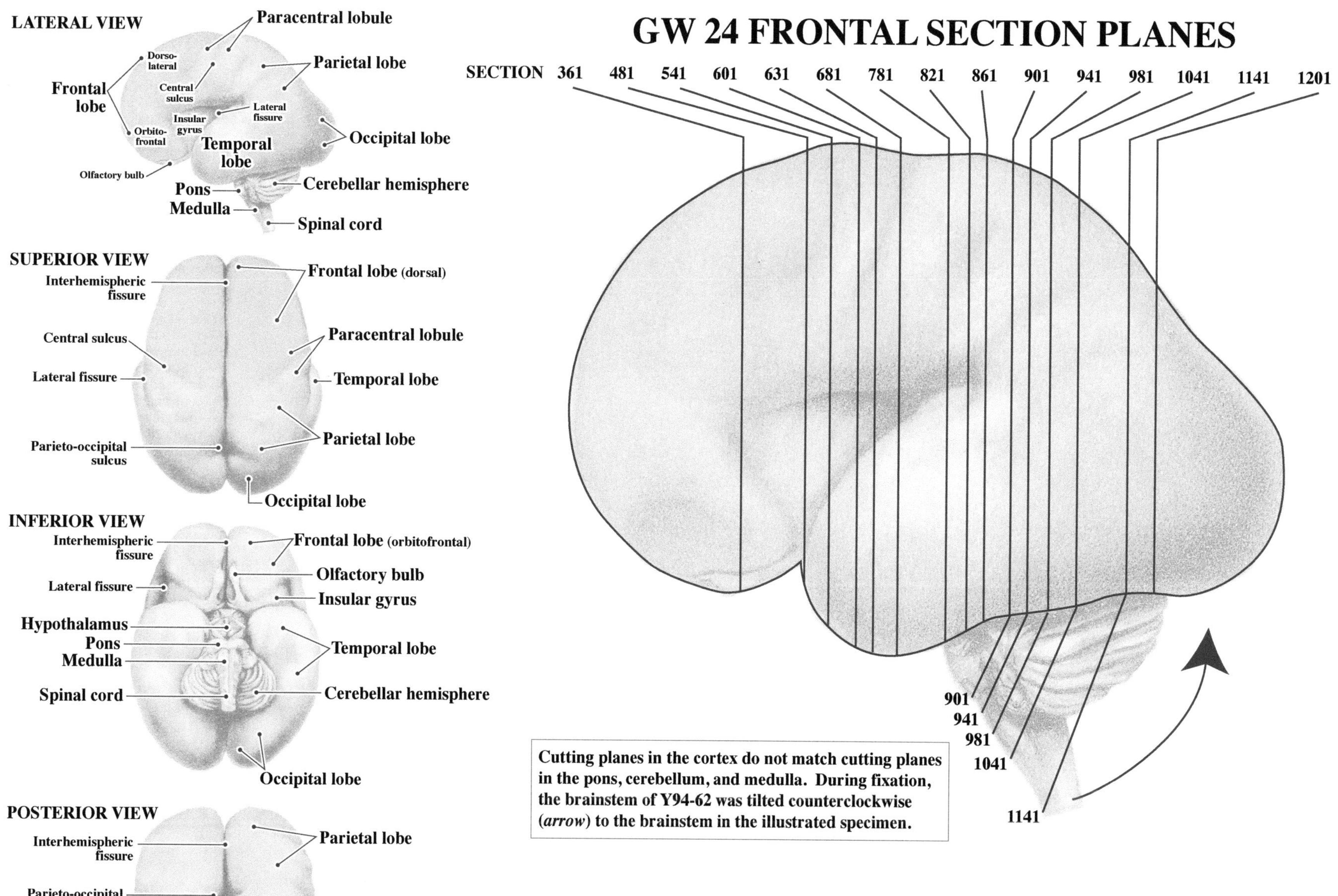

Figure 5. The *column on the left* shows an approximate GW 24 brain from lateral, superior, inferior, and posterior views with major structures labeled (Figures 1 to 4 in Table 11, Volume 2, Retzius, 1896). An enlarged lateral view with the approximate locations and cutting angles of the sections of Y94-62 is shown *above*. Note that the cutting angles in the brainstem do not match the cutting angles in the cortex. After dissection, the brainstem has no support to maintain a constant angle of downward extension from the cortex and can easily be flexed either forward or backward during fixation. In Y94-62, the brainstem was flexed counterclockwise (*arrow*) to tuck under the cerebral hemispheres.

PLATE 31A
CR 210 mm
GW 24, Y94-62
Frontal
Section 361

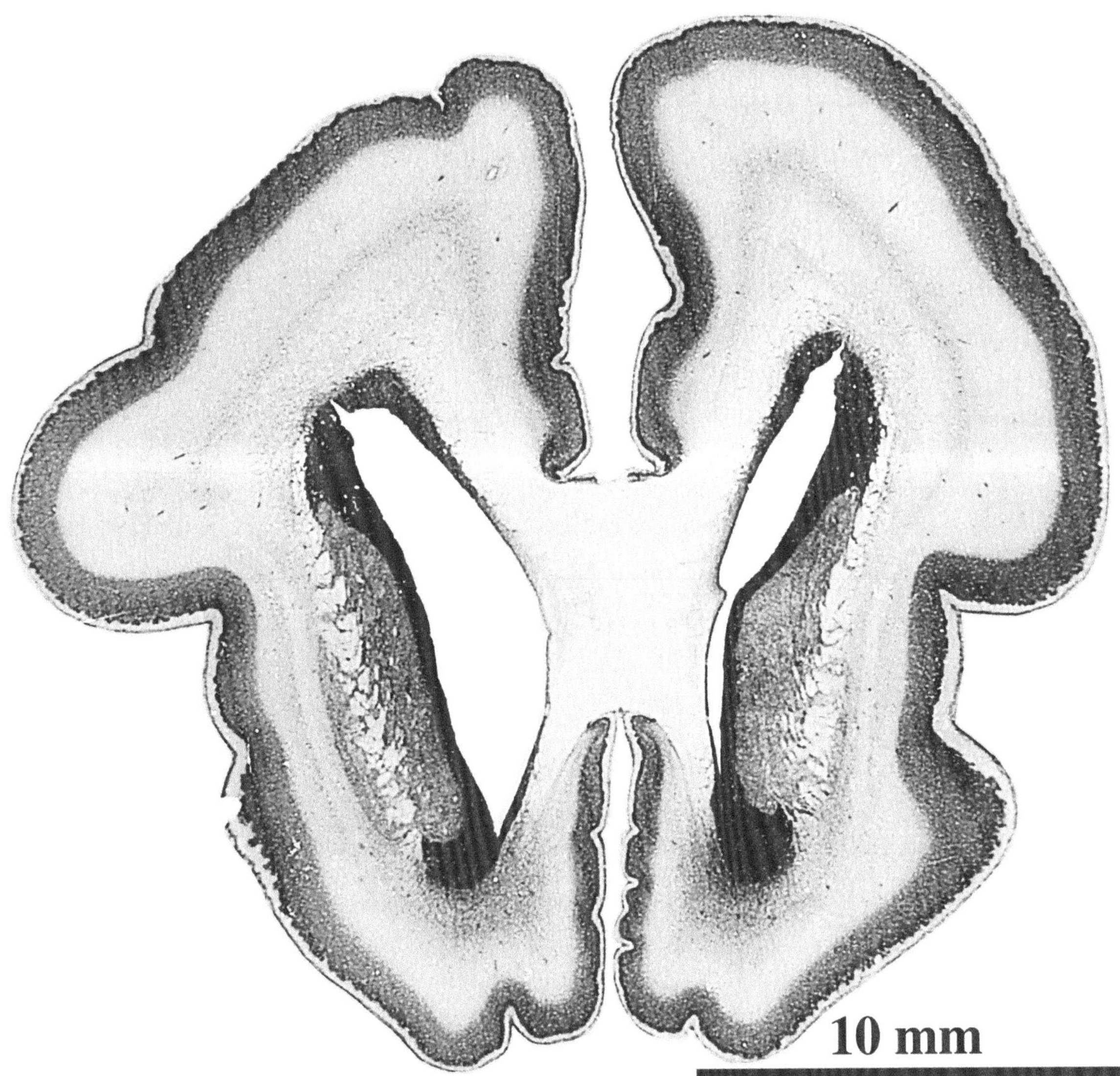

LAYERS OF THE CORTICAL *STRATIFIED TRANSITIONAL FIELD (STF)*

STF1*—**Superficial fibrous layer with an early developmental stage ***(t1) when many cells are migrating through it, followed by a late stage ***(t2)*** with sparse cells. Endures as the subcortical white matter.

***STF2*—**Upper cellular layer, the last sojourn zone before cells translocate to the cortical plate.

***STF4*—**Complex middle layer with three developmental stages:
t1– fibrous layer without interspersed cells;
t2– cells and fibers intermingle to form striations;
t3– fibers endure in the deep white matter.

***STF5*—**Deep cellular layer, the first sojourn zone to appear outside the germinal matrix.

***STF6*—**Late-forming deep layer of callosal fibers outside the germinal matrix.

PLATE 31B

GEP - glioepithelium
NEP - neuroepithelium
SVZ - subventricular zone
Germinal and transitional structures in *italics*

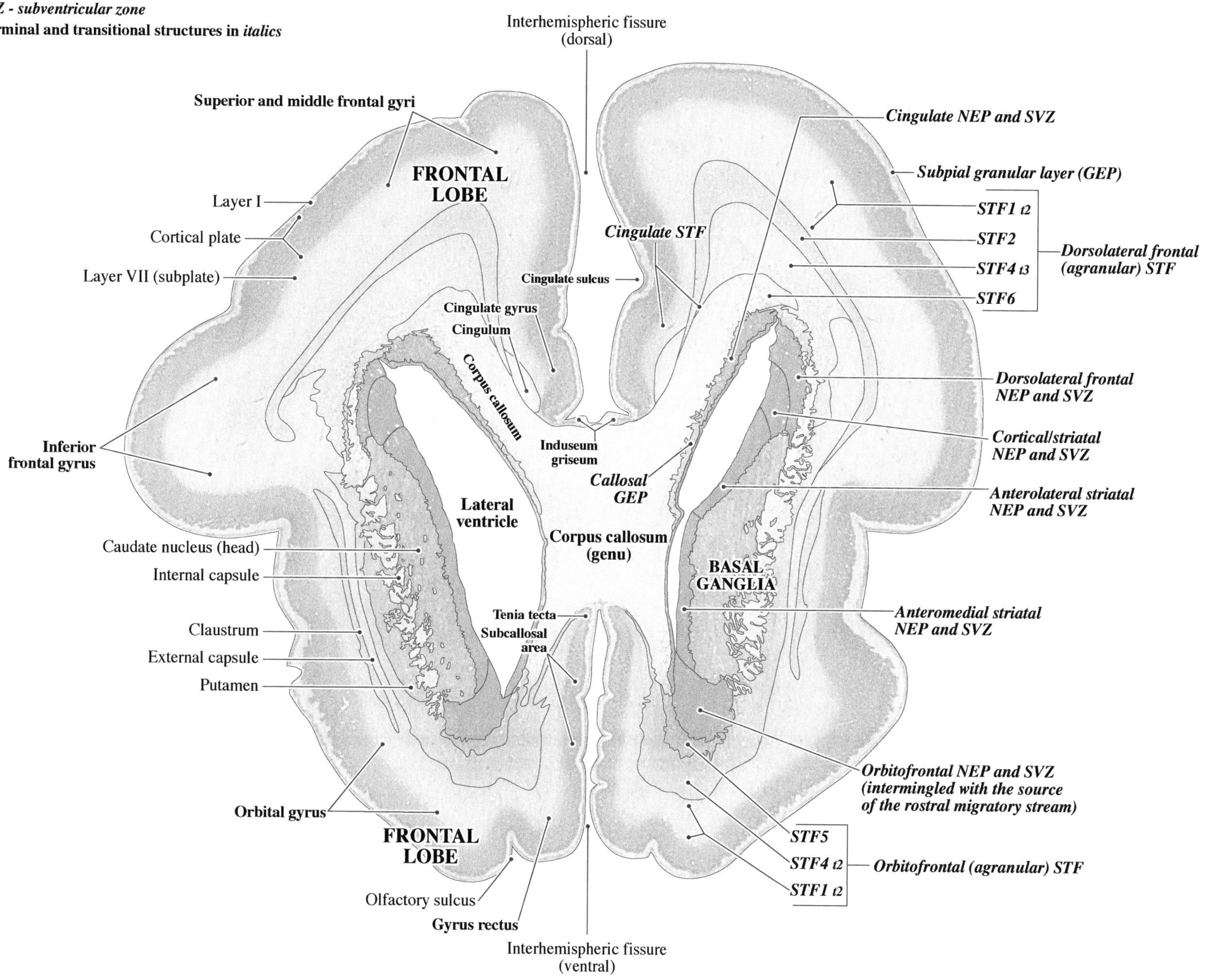

Interhemispheric fissure (dorsal)

Superior and middle frontal gyri

FRONTAL LOBE

Layer I

Cortical plate

Layer VII (subplate)

Cingulate STF

Cingulate sulcus

Cingulate gyrus

Cingulum

Corpus callosum

Induseum griseum

Callosal GEP

Inferior frontal gyrus

Lateral ventricle

Corpus callosum (genu)

BASAL GANGLIA

Caudate nucleus (head)

Internal capsule

Tenia tecta

Subcallosal area

Claustrum

External capsule

Putamen

Orbital gyrus

FRONTAL LOBE

Olfactory sulcus

Gyrus rectus

Interhemispheric fissure (ventral)

Cingulate NEP and SVZ

Subpial granular layer (GEP)

STF1 t2

STF2

STF4 t3

STF6

Dorsolateral frontal (agranular) STF

Dorsolateral frontal NEP and SVZ

Cortical/striatal NEP and SVZ

Anterolateral striatal NEP and SVZ

Anteromedial striatal NEP and SVZ

Orbitofrontal NEP and SVZ (intermingled with the source of the rostral migratory stream)

STF5

STF4 t2

STF1 t2

Orbitofrontal (agranular) STF

PLATE 32A
CR 210 mm
GW 24, Y94-62
Frontal
Section 481

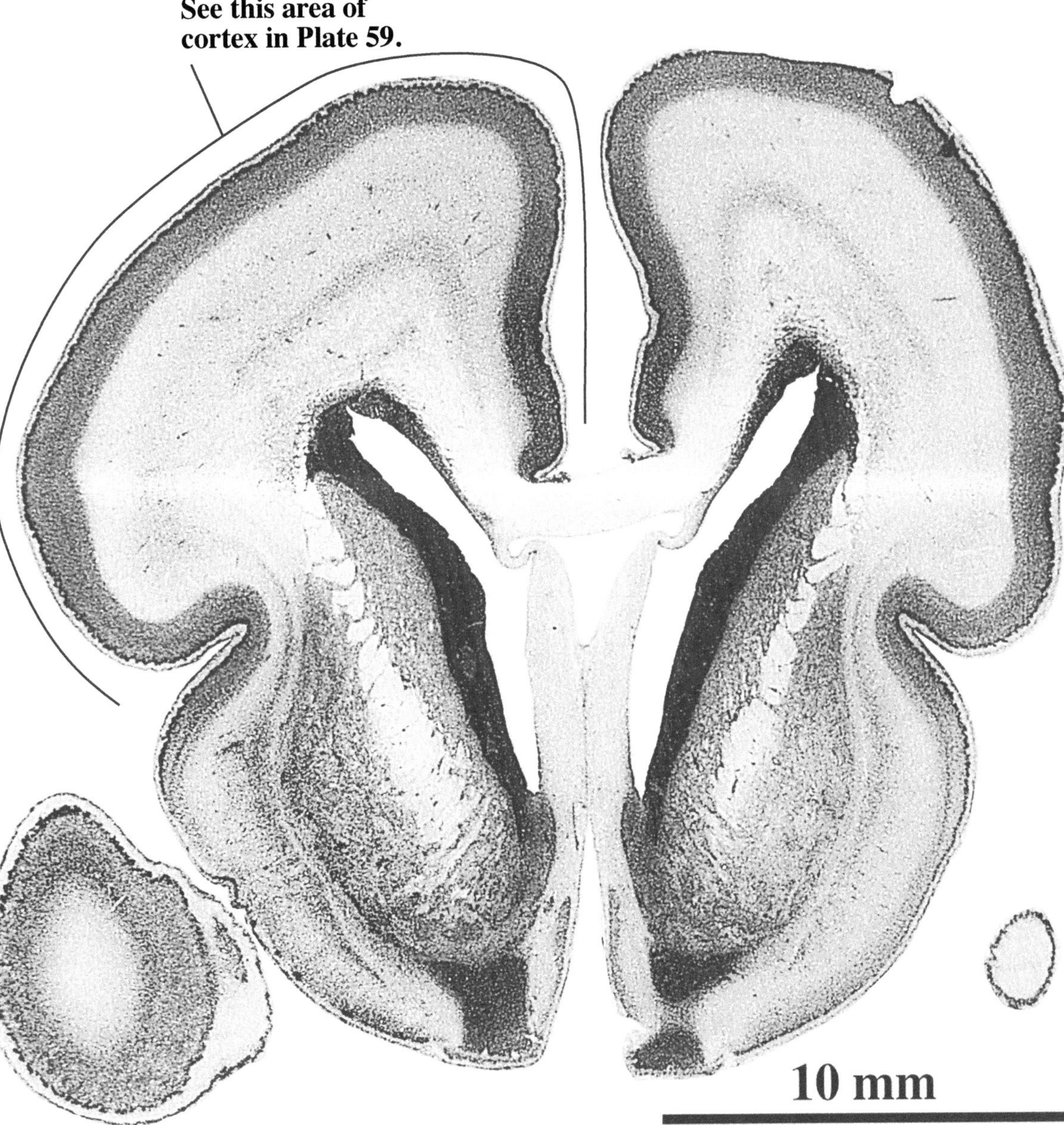

LAYERS OF THE CORTICAL *STRATIFIED TRANSITIONAL FIELD (STF)*

STF1—Superficial fibrous layer with an early developmental stage ***(t1)*** when many cells are migrating through it, followed by a late stage ***(t2)*** with sparse cells. Endures as the subcortical white matter.

STF2—Upper cellular layer, the last sojourn zone before cells translocate to the cortical plate.

STF4—Complex middle layer with three developmental stages:
t1– fibrous layer without interspersed cells;
t2– cells and fibers intermingle to form striations;
t3– fibers endure in the deep white matter.

STF6—Late-forming deep layer of callosal fibers outside the germinal matrix.

PLATE 32B

G/EP - glioepithelium/ependyma
GEP - glioepithelium
NEP - neuroepithelium
SVZ - subventricular zone
Germinal and transitional structures in *italics*

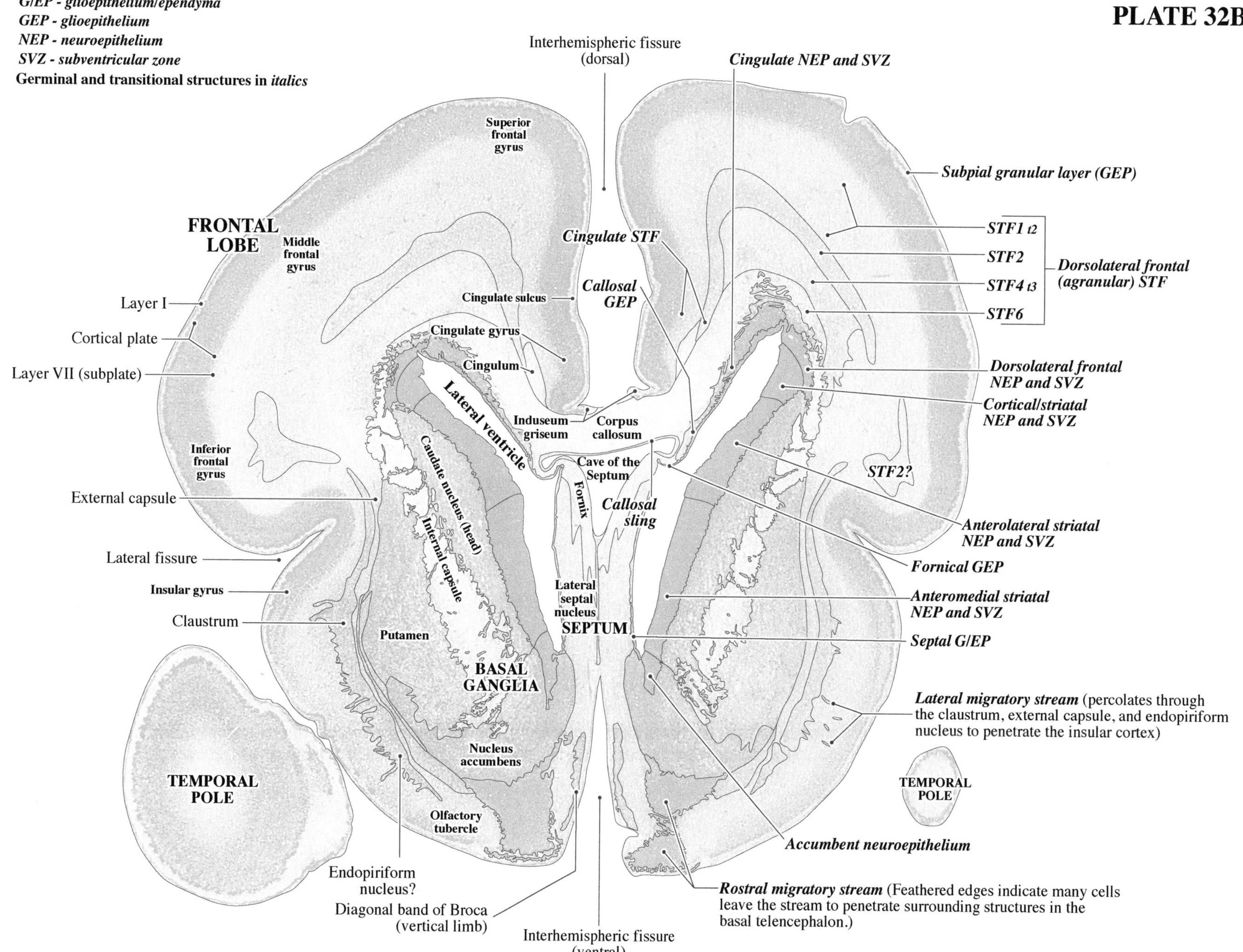

PLATE 33A
CR 210 mm
GW 24, Y94-62
Frontal
Section 541

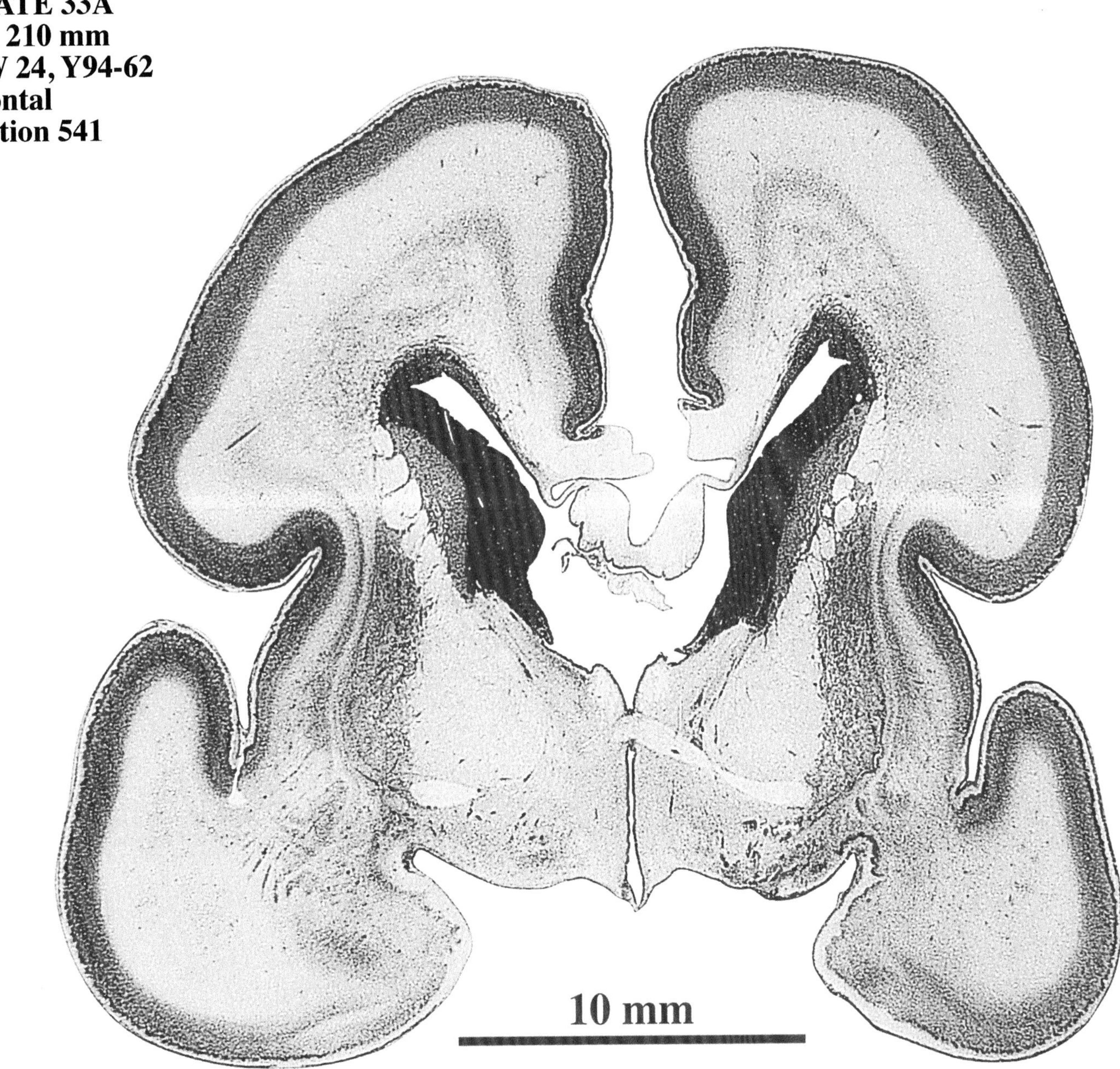

LAYERS OF THE CORTICAL *STRATIFIED TRANSITIONAL FIELD (STF)*

STF1*—**Superficial fibrous layer with an early developmental stage ***(t1) when many cells are migrating through it, followed by a late stage ***(t2)*** with sparse cells. Endures as the subcortical white matter.

***STF2*—**Upper cellular layer, the last sojourn zone before cells translocate to the cortical plate.

***STF4*—**Complex middle layer with three developmental stages:
t1– fibrous layer without interspersed cells;
t2– cells and fibers intermingle to form striations;
t3– fibers endure in the deep white matter.

***STF6*—**Late-forming deep layer of callosal fibers outside the germinal matrix.

See detail of the brain core in Plates 46A and B.

PLATE 33B

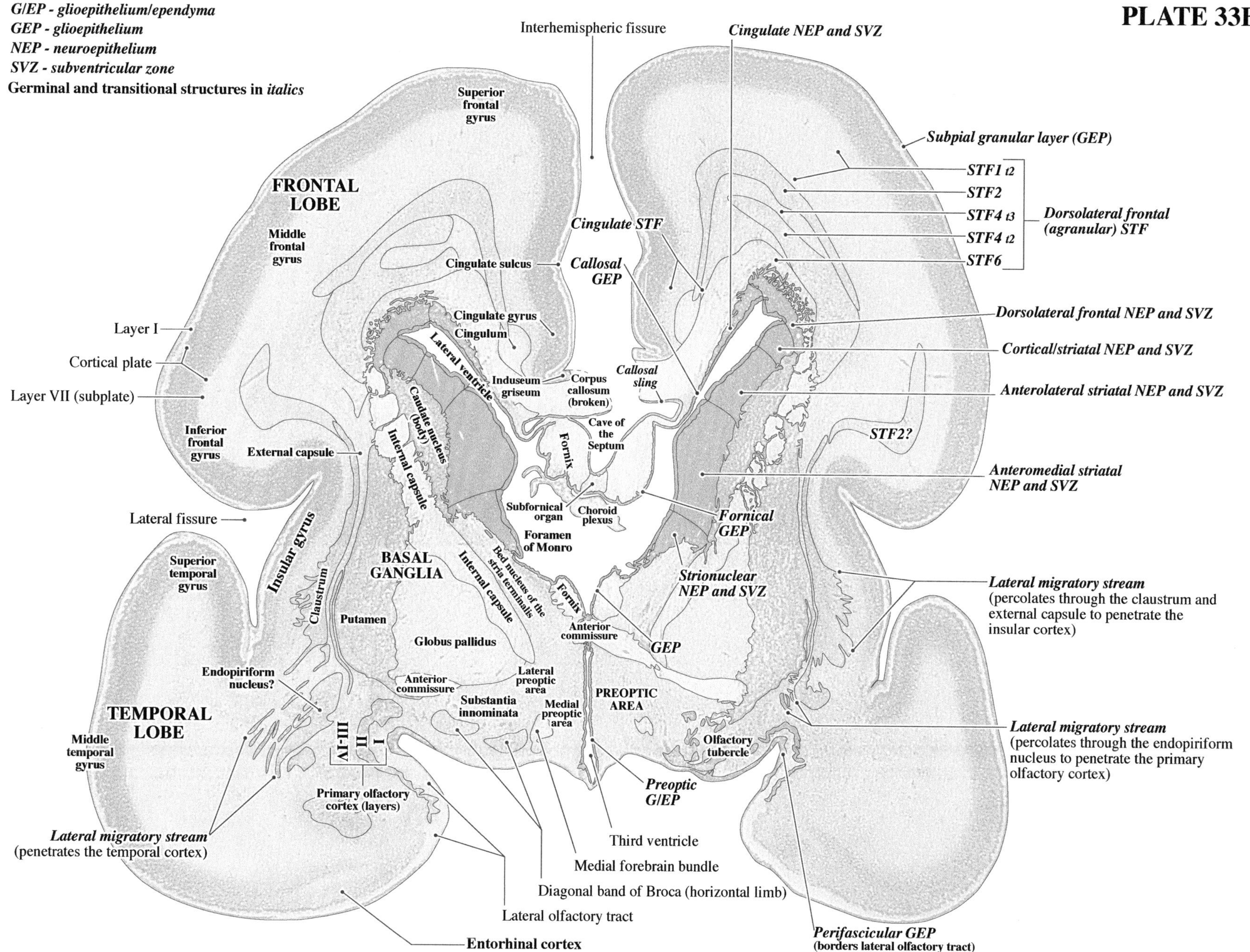
G/EP - glioepithelium/ependyma
GEP - glioepithelium
NEP - neuroepithelium
SVZ - subventricular zone
Germinal and transitional structures in *italics*
Interhemispheric fissure
Cingulate NEP and SVZ
Superior frontal gyrus
Subpial granular layer (GEP)
STF1 t2
STF2
STF4 t3
STF4 t2
STF6
Dorsolateral frontal (agranular) STF
FRONTAL LOBE
Middle frontal gyrus
Cingulate STF
Cingulate sulcus
Callosal GEP
Cingulate gyrus
Cingulum
Dorsolateral frontal NEP and SVZ
Layer I
Cortical plate
Layer VII (subplate)
Lateral ventricle
Cortical/striatal NEP and SVZ
Induseum griseum
Corpus callosum (broken)
Callosal sling
Anterolateral striatal NEP and SVZ
Caudate nucleus (body)
Inferior frontal gyrus
External capsule
Internal capsule
Fornix
Cave of the Septum
STF2?
Anteromedial striatal NEP and SVZ
Subfornical organ
Choroid plexus
Fornical GEP
Lateral fissure
Insular gyrus
Foramen of Monro
Superior temporal gyrus
BASAL GANGLIA
Claustrum
Bed nucleus of the stria terminalis
Internal capsule
Fornix
Strionuclear NEP and SVZ
Lateral migratory stream
(percolates through the claustrum and external capsule to penetrate the insular cortex)
Putamen
Globus pallidus
Anterior commissure
GEP
Endopiriform nucleus?
Anterior commissure
Lateral preoptic area
Substantia innominata
Medial preoptic area
PREOPTIC AREA
TEMPORAL LOBE
Middle temporal gyrus
III-IV
II
I
Olfactory tubercle
Lateral migratory stream
(percolates through the endopiriform nucleus to penetrate the primary olfactory cortex)
Primary olfactory cortex (layers)
Preoptic G/EP
Lateral migratory stream
(penetrates the temporal cortex)
Third ventricle
Medial forebrain bundle
Diagonal band of Broca (horizontal limb)
Lateral olfactory tract
Entorhinal cortex
Perifascicular GEP
(borders lateral olfactory tract)

PLATE 34A
CR 210 mm
GW 24, Y94-62
Frontal
Section 601

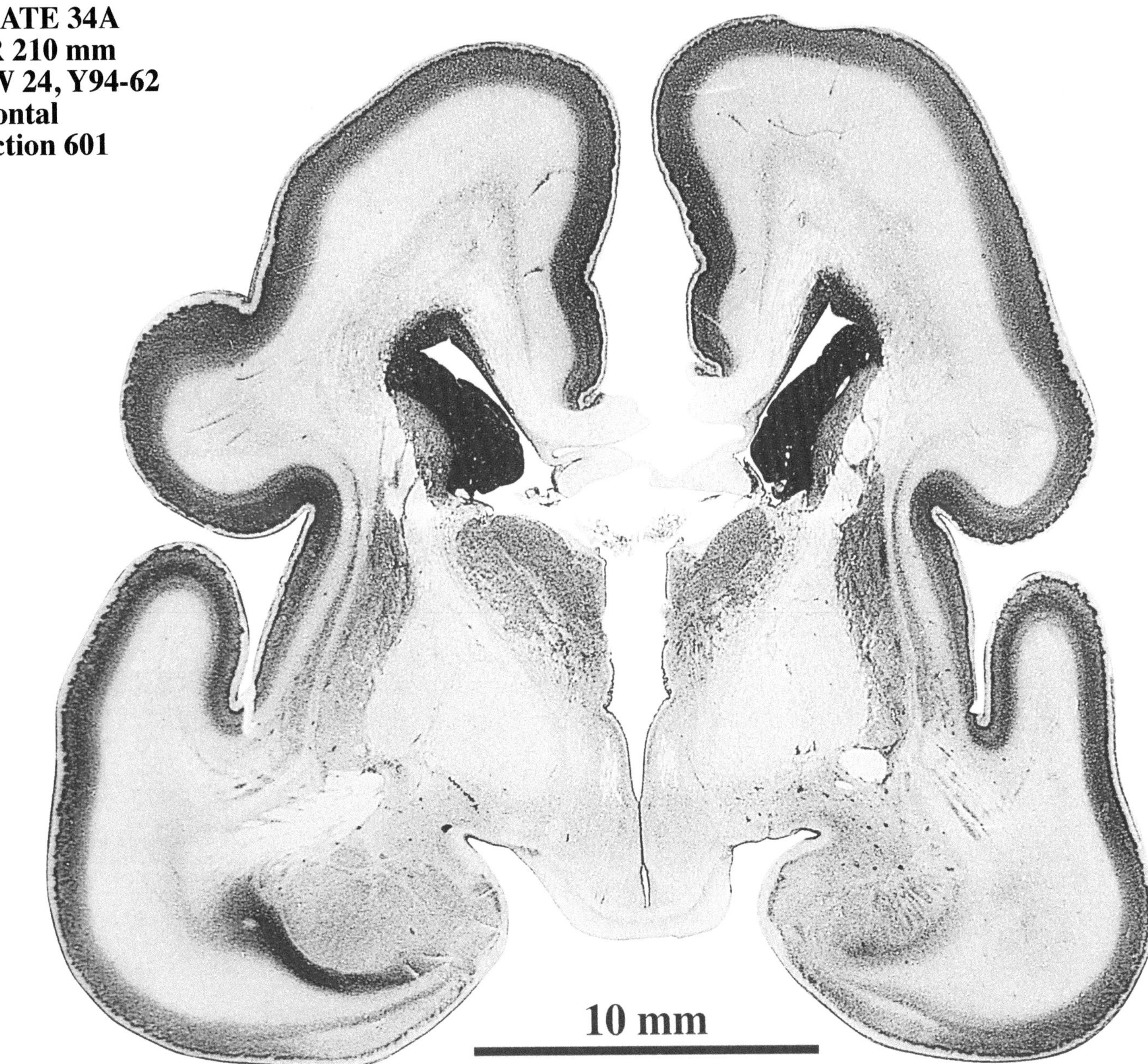

LAYERS OF THE CORTICAL *STRATIFIED TRANSITIONAL* FIELD (STF)

STF1—Superficial fibrous layer with an early developmental stage ***(t1)*** when many cells are migrating through it, followed by a late stage ***(t2)*** with sparse cells. Endures as the subcortical white matter.

STF2—Upper cellular layer, the last sojourn zone before cells translocate to the cortical plate.

STF4—Complex middle layer with three developmental stages:
t1– fibrous layer without interspersed cells;
t2– cells and fibers intermingle to form striations;
t3– fibers endure in the deep white matter.

STF5—Deep cellular layer, the first sojourn zone to appear outside the germinal matrix.

STF6—Late-forming deep layer of callosal fibers outside the germinal matrix.

See detail of the brain core in Plates 47A and B.

PLATE 34B

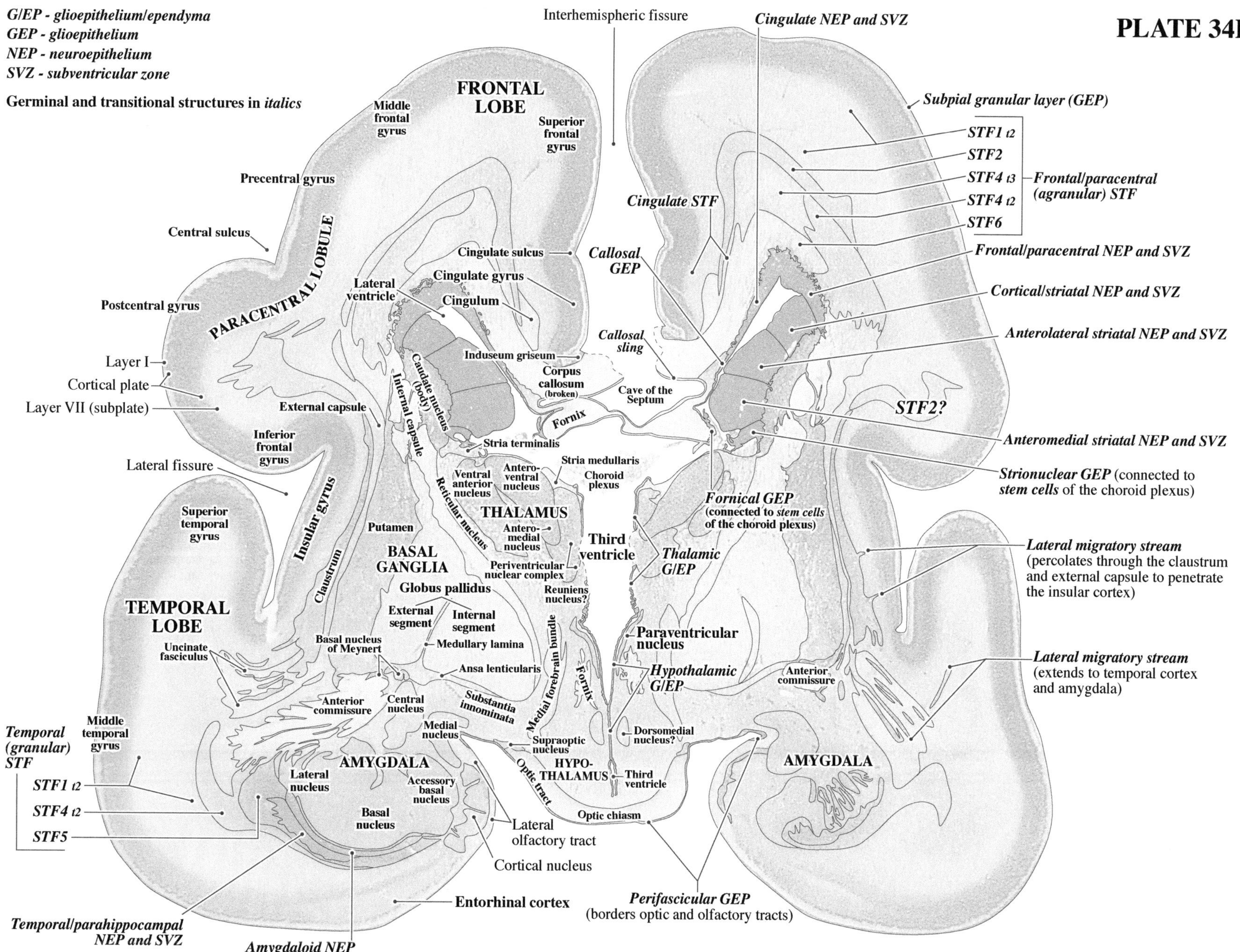

G/EP - glioepithelium/ependyma
GEP - glioepithelium
NEP - neuroepithelium
SVZ - subventricular zone
Germinal and transitional structures in italics
Interhemispheric fissure
Cingulate NEP and SVZ
FRONTAL LOBE
Middle frontal gyrus
Superior frontal gyrus
Subpial granular layer (GEP)
STF1 t2
STF2
STF4 t3
STF4 t2
STF6
Frontal/paracentral (agranular) STF
Precentral gyrus
Cingulate STF
Central sulcus
Frontal/paracentral NEP and SVZ
Cingulate sulcus
Callosal GEP
Lateral ventricle
Cingulate gyrus
Cortical/striatal NEP and SVZ
Postcentral gyrus
PARACENTRAL LOBULE
Cingulum
Callosal sling
Anterolateral striatal NEP and SVZ
Induseum griseum
Layer I
Corpus callosum (broken)
Cave of the Septum
Cortical plate
Caudate nucleus (body)
Internal capsule
STF2?
Layer VII (subplate)
External capsule
Fornix
Anteromedial striatal NEP and SVZ
Inferior frontal gyrus
Stria terminalis
Stria medullaris
Lateral fissure
Strionuclear GEP (connected to stem cells of the choroid plexus)
Ventral anterior nucleus
Antero-ventral nucleus
Choroid plexus
Reticular nucleus
Fornical GEP (connected to stem cells of the choroid plexus)
Superior temporal gyrus
THALAMUS
Insular gyrus
Putamen
Antero-medial nucleus
Third ventricle
Thalamic G/EP
BASAL GANGLIA
Periventricular nuclear complex
Lateral migratory stream (percolates through the claustrum and external capsule to penetrate the insular cortex)
Claustrum
Globus pallidus
Reuniens nucleus?
TEMPORAL LOBE
External segment
Internal segment
Paraventricular nucleus
Uncinate fasciculus
Basal nucleus of Meynert
Medullary lamina
Medial forebrain bundle
Ansa lenticularis
Fornix
Hypothalamic G/EP
Anterior commissure
Lateral migratory stream (extends to temporal cortex and amygdala)
Anterior commissure
Central nucleus
Substantia innominata
Middle temporal gyrus
Medial nucleus
Dorsomedial nucleus?
Temporal (granular) STF
Supraoptic nucleus
AMYGDALA
HYPO-THALAMUS
Third ventricle
AMYGDALA
STF1 t2
Lateral nucleus
Accessory basal nucleus
Optic tract
STF4 t2
Basal nucleus
Optic chiasm
STF5
Lateral olfactory tract
Cortical nucleus
Entorhinal cortex
Perifascicular GEP (borders optic and olfactory tracts)
Temporal/parahippocampal NEP and SVZ
Amygdaloid NEP

PLATE 35A
CR 210 mm
GW 24, Y94-62
Frontal
Section 631

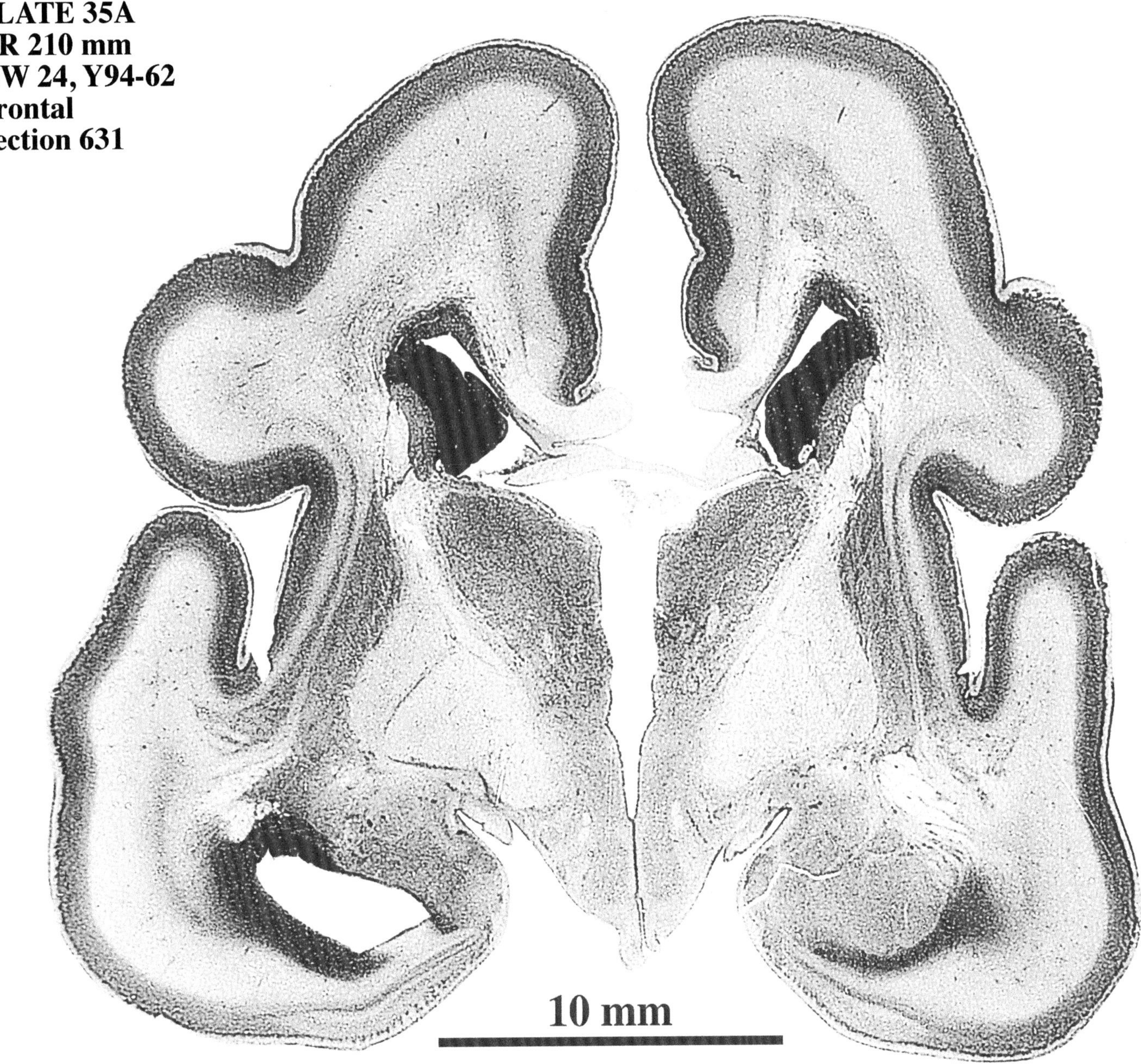

LAYERS OF THE CORTICAL *STRATIFIED TRANSITIONAL FIELD (STF)*

STF1—Superficial fibrous layer with an early developmental stage ***(t1)*** when many cells are migrating through it, followed by a late stage ***(t2)*** with sparse cells. Endures as the subcortical white matter.

STF2—Upper cellular layer, the last sojourn zone before cells translocate to the cortical plate.

STF3—Honeycomb trilaminar matrix ***(3a, 3b, 3c)*** of cells and fibers found only in granular cortices.

STF4—Complex middle layer with three developmental stages:
t1– fibrous layer without interspersed cells;
t2– cells and fibers intermingle to form striations;
t3– fibers endure in the deep white matter.

STF5—Deep cellular layer, the first sojourn zone to appear outside the germinal matrix.

STF6—Late-forming deep layer of callosal fibers outside the germinal matrix.

See detail of brain core in Plates 48A and B.

PLATE 35B

G/EP - glioepithelium/ependyma
GEP - glioepithelium
NEP - neuroepithelium
SVZ - subventricular zone

Germinal and transitional structures in *italics*

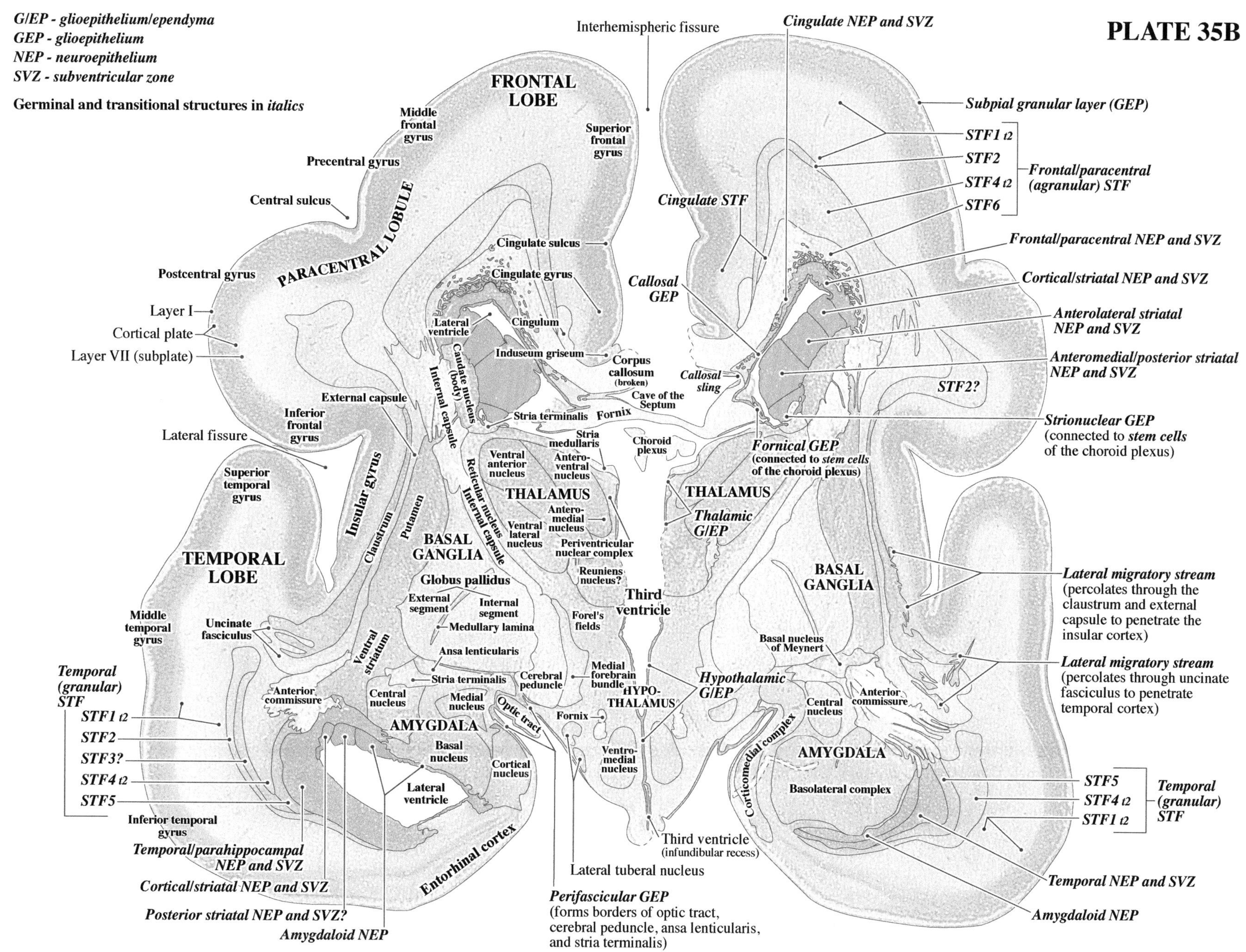

PLATE 36A
CR 210 mm
GW 24, Y94-62
Frontal
Section 681

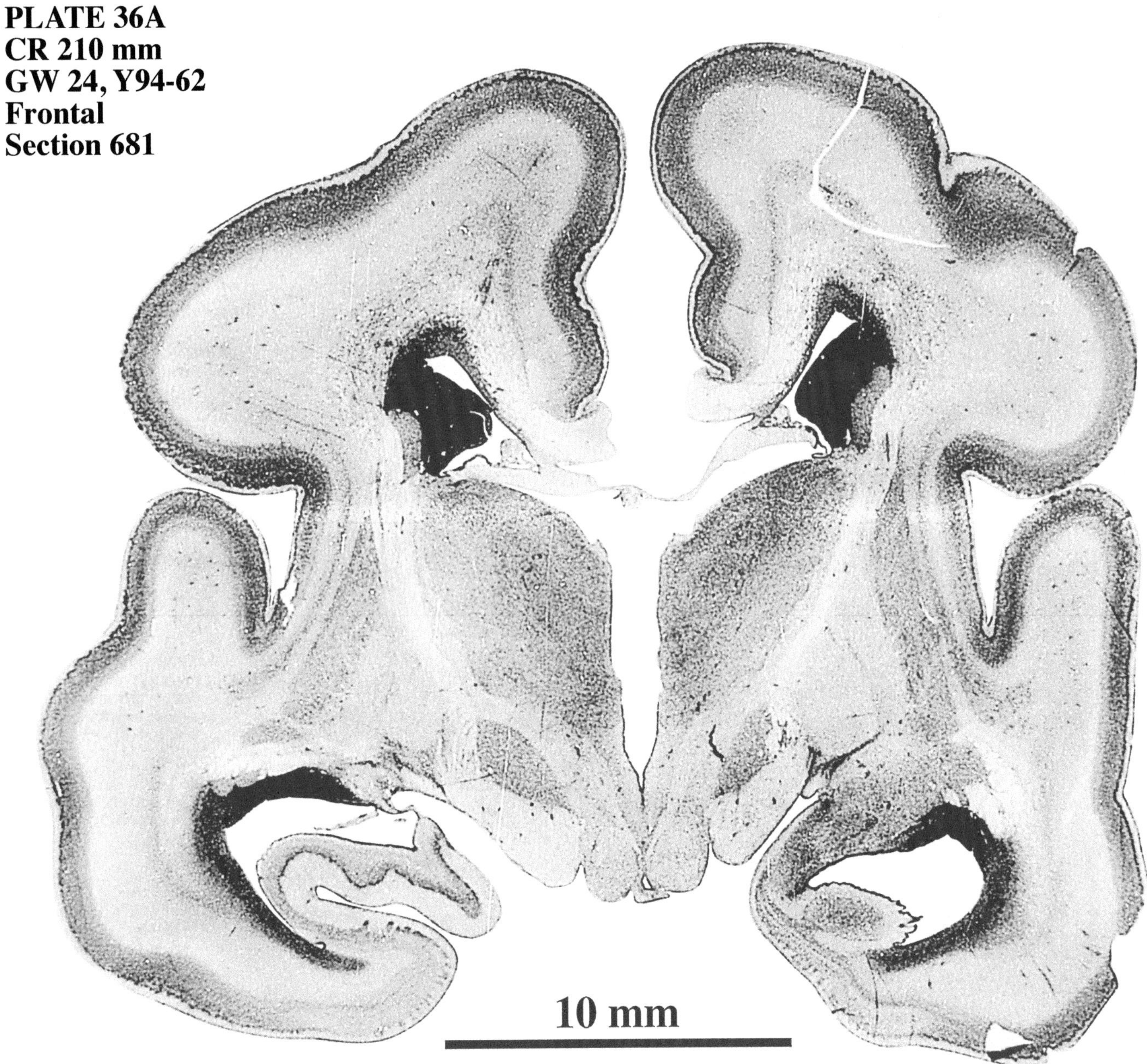

LAYERS OF THE CORTICAL *STRATIFIED TRANSITIONAL FIELD (STF)*

STF1—Superficial fibrous layer with an early developmental stage ***(t1)*** when many cells are migrating through it, followed by a late stage ***(t2)*** with sparse cells. Endures as the subcortical white matter.

STF2—Upper cellular layer, the last sojourn zone before cells translocate to the cortical plate.

STF3—Honeycomb trilaminar matrix ***(3a, 3b, 3c)*** of cells and fibers found only in granular cortices.

STF4—Complex middle layer with three developmental stages:
t1– fibrous layer without interspersed cells;
t2– cells and fibers intermingle to form striations;
t3– fibers endure in the deep white matter.

STF5—Deep cellular layer, the first sojourn zone to appear outside the germinal matrix.

STF6—Late-forming deep layer of callosal fibers outside the germinal matrix.

See detail of brain core in Plates 49A and B.

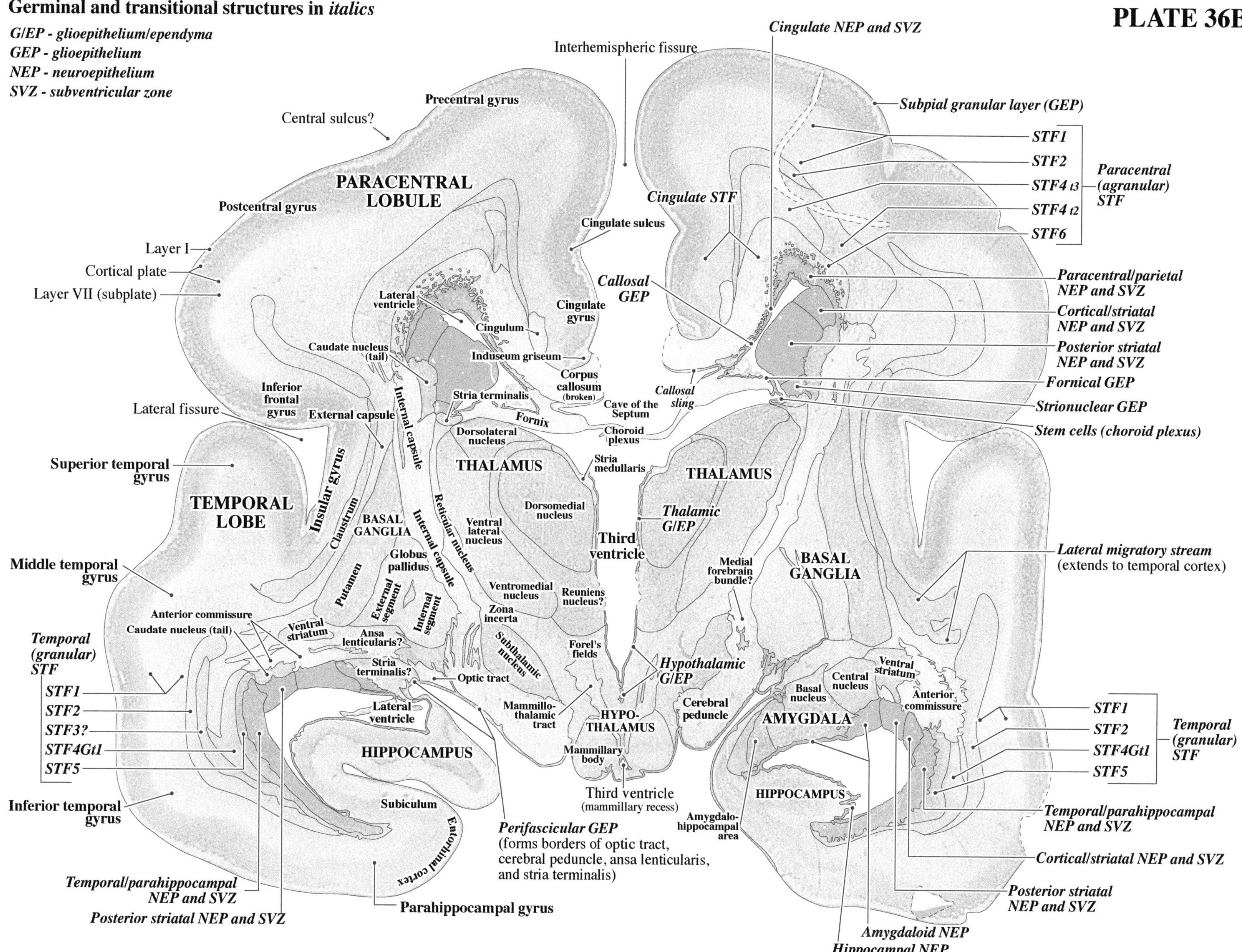
Germinal and transitional structures in italics
G/EP - glioepithelium/ependyma
GEP - glioepithelium
NEP - neuroepithelium
SVZ - subventricular zone
Interhemispheric fissure
Cingulate NEP and SVZ
Precentral gyrus
Central sulcus?
PARACENTRAL LOBULE
Postcentral gyrus
Layer I
Cortical plate
Layer VII (subplate)
Cingulate sulcus
Cingulate STF
Callosal GEP
Subpial granular layer (GEP)
STF1
STF2
STF4 t3
STF4 t2
STF6
Paracentral (agranular) STF
Paracentral/parietal NEP and SVZ
Cortical/striatal NEP and SVZ
Posterior striatal NEP and SVZ
Fornical GEP
Strionuclear GEP
Stem cells (choroid plexus)
Lateral ventricle
Cingulum
Cingulate gyrus
Caudate nucleus (tail)
Induseum griseum
Corpus callosum (broken)
Cave of the Septum
Callosal sling
Inferior frontal gyrus
Stria terminalis
Fornix
Choroid plexus
Lateral fissure
External capsule
Internal capsule
Dorsolateral nucleus
Stria medullaris
Superior temporal gyrus
THALAMUS
TEMPORAL LOBE
Insular gyrus
Claustrum
BASAL GANGLIA
Reticular nucleus
Ventral lateral nucleus
Dorsomedial nucleus
Third ventricle
Thalamic G/EP
Middle temporal gyrus
Globus pallidus
Putamen
External segment
Internal segment
Ventromedial nucleus
Reuniens nucleus?
Medial forebrain bundle?
Lateral migratory stream (extends to temporal cortex)
Anterior commissure
Caudate nucleus (tail)
Ventral striatum
Ansa lenticularis?
Zona incerta
Subthalamic nucleus
Forel's fields
Hypothalamic G/EP
Temporal (granular) STF
STF1
STF2
STF3?
STF4Gt1
STF5
Stria terminalis?
Optic tract
Lateral ventricle
Mammillo-thalamic tract
HYPO-THALAMUS
Cerebral peduncle
Basal nucleus
Central nucleus
Ventral striatum
Anterior commissure
AMYGDALA
STF1
STF2
STF4Gt1
STF5
Temporal (granular) STF
HIPPOCAMPUS
Mammillary body
Inferior temporal gyrus
Subiculum
Third ventricle (mammillary recess)
HIPPOCAMPUS
Amygdalo-hippocampal area
Temporal/parahippocampal NEP and SVZ
Entorhinal cortex
Perifascicular GEP (forms borders of optic tract, cerebral peduncle, ansa lenticularis, and stria terminalis)
Cortical/striatal NEP and SVZ
Temporal/parahippocampal NEP and SVZ
Posterior striatal NEP and SVZ
Parahippocampal gyrus
Posterior striatal NEP and SVZ
Amygdaloid NEP
Hippocampal NEP

PLATE 37A
CR 210 mm
GW 24, Y94-62
Frontal
Section 781

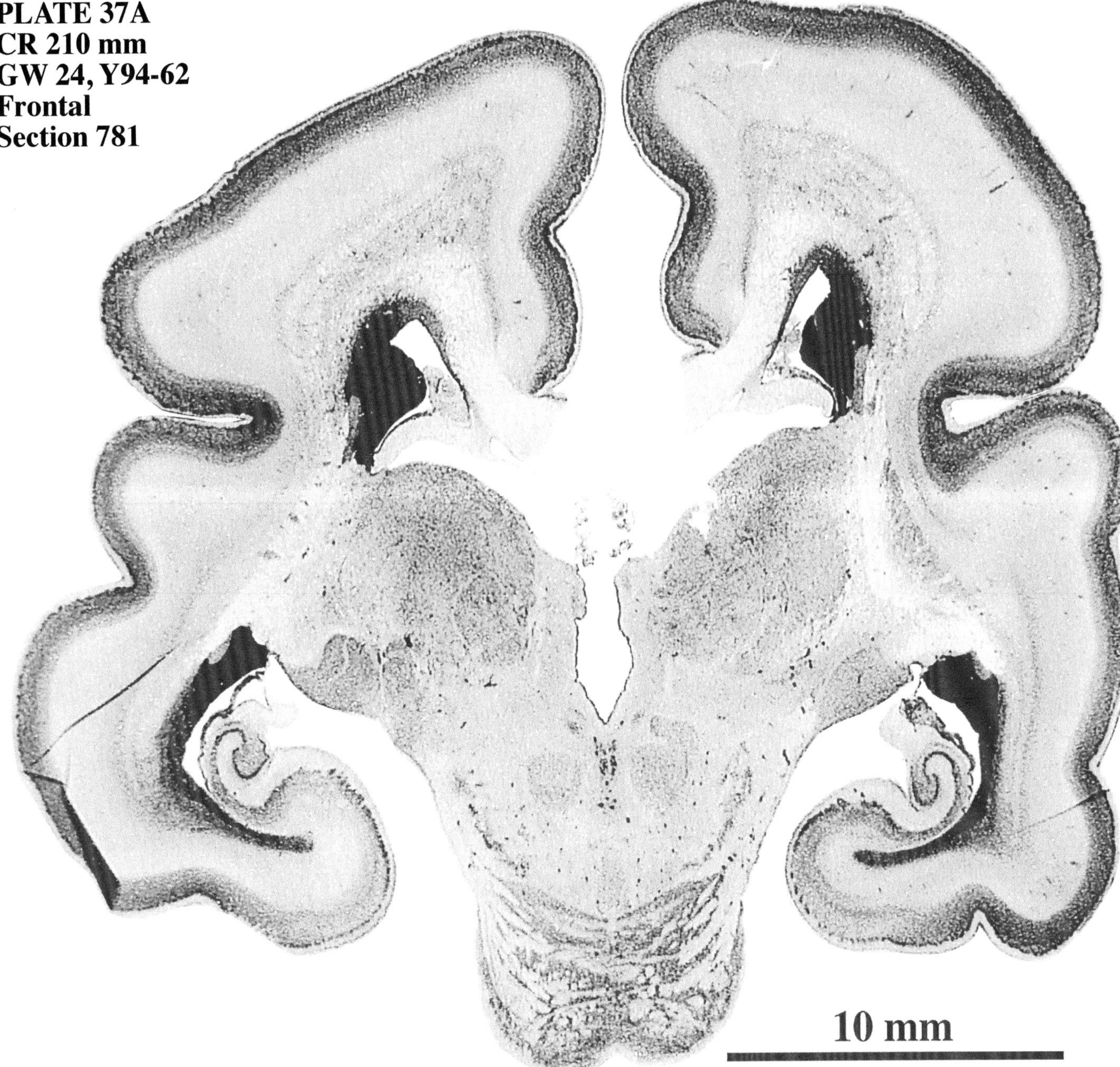

LAYERS OF THE CORTICAL *STRATIFIED TRANSITIONAL FIELD (STF)*

STF1—Superficial fibrous layer with an early developmental stage ***(t1)*** when many cells are migrating through it, followed by a late stage ***(t2)*** with sparse cells. Endures as the subcortical white matter.

STF2—Upper cellular layer, the last sojourn zone before cells translocate to the cortical plate.

STF3—Honeycomb trilaminar matrix ***(3a, 3b, 3c)*** of cells and fibers found only in granular cortices.

STF4—Complex middle layer with three developmental stages:
t1– fibrous layer without interspersed cells;
t2– cells and fibers intermingle to form striations;
t3– fibers endure in the deep white matter.

STF5—Deep cellular layer, the first sojourn zone to appear outside the germinal matrix.

STF6—Late-forming deep layer of callosal fibers outside the germinal matrix.

See detail of brain core in Plates 50A and B.

PLATE 37B

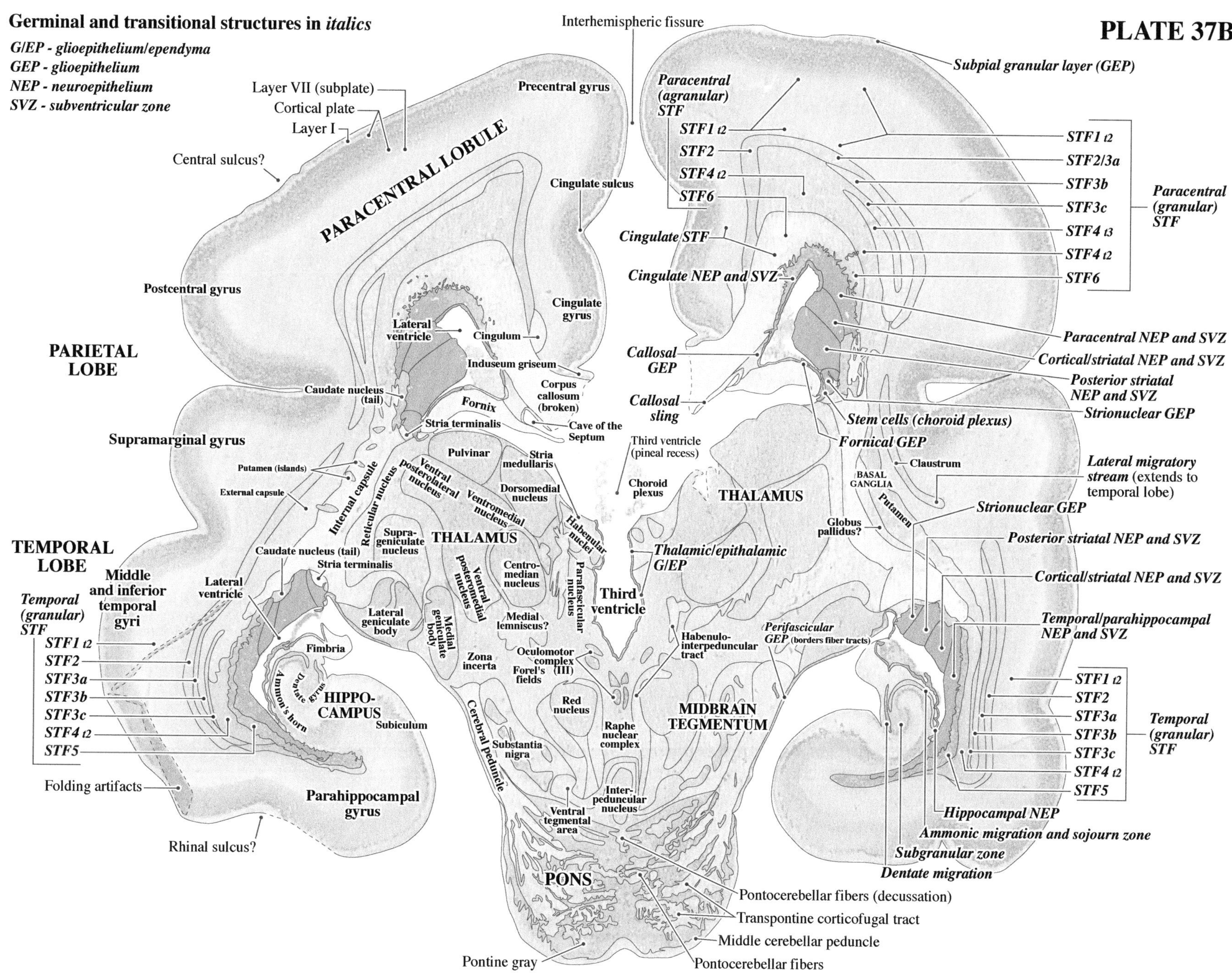

Germinal and transitional structures in *italics*
G/EP - glioepithelium/ependyma
GEP - glioepithelium
NEP - neuroepithelium
SVZ - subventricular zone
Interhemispheric fissure
Subpial granular layer (GEP)
Layer VII (subplate)
Cortical plate
Layer I
Central sulcus?
Precentral gyrus
PARACENTRAL LOBULE
Paracentral (agranular) STF
STF1 t2
STF2
STF4 t2
STF6
STF2/3a
STF3b
STF3c
STF4 t3
Paracentral (granular) STF
Cingulate sulcus
Cingulate STF
Cingulate NEP and SVZ
Postcentral gyrus
Cingulate gyrus
Lateral ventricle
Cingulum
Paracentral NEP and SVZ
Cortical/striatal NEP and SVZ
PARIETAL LOBE
Callosal GEP
Induseum griseum
Corpus callosum (broken)
Posterior striatal NEP and SVZ
Caudate nucleus (tail)
Fornix
Callosal sling
Strionuclear GEP
Stria terminalis
Cave of the Septum
Stem cells (choroid plexus)
Supramarginal gyrus
Third ventricle (pineal recess)
Fornical GEP
Pulvinar
Stria medullaris
Claustrum
Putamen (islands)
Ventral posterolateral nucleus
Lateral migratory stream (extends to temporal lobe)
Internal capsule
Reticular nucleus
Ventromedial nucleus
Dorsomedial nucleus
Choroid plexus
BASAL GANGLIA
External capsule
THALAMUS
Putamen
Habenular nuclei
Globus pallidus?
Supra-geniculate nucleus
TEMPORAL LOBE
Caudate nucleus (tail)
Thalamic/epithalamic G/EP
Middle and inferior temporal gyri
Stria terminalis
Centro-median nucleus
Ventral posteromedial nucleus
Parafascicular nucleus
Lateral ventricle
Third ventricle
Temporal (granular) STF
Lateral geniculate body
Medial geniculate body
Medial lemniscus?
Temporal/parahippocampal NEP and SVZ
STF1 t2
Habenulo-interpeduncular tract
Perifascicular GEP (borders fiber tracts)
STF2
Fimbria
Zona incerta
Oculomotor complex (III)
Forel's fields
STF3a
Dentate gyrus
Ammon's horn
HIPPO-CAMPUS
STF3b
Red nucleus
MIDBRAIN TEGMENTUM
STF3c
Subiculum
Cerebral peduncle
Raphe nuclear complex
STF4 t2
Substantia nigra
STF5
Folding artifacts
Parahippocampal gyrus
Inter-peduncular nucleus
Ventral tegmental area
Hippocampal NEP
Ammonic migration and sojourn zone
Rhinal sulcus?
Subgranular zone
PONS
Dentate migration
Pontocerebellar fibers (decussation)
Transpontine corticofugal tract
Middle cerebellar peduncle
Pontine gray
Pontocerebellar fibers

PLATE 38A
CR 210 mm
GW 24, Y94-62
Frontal
Section 821

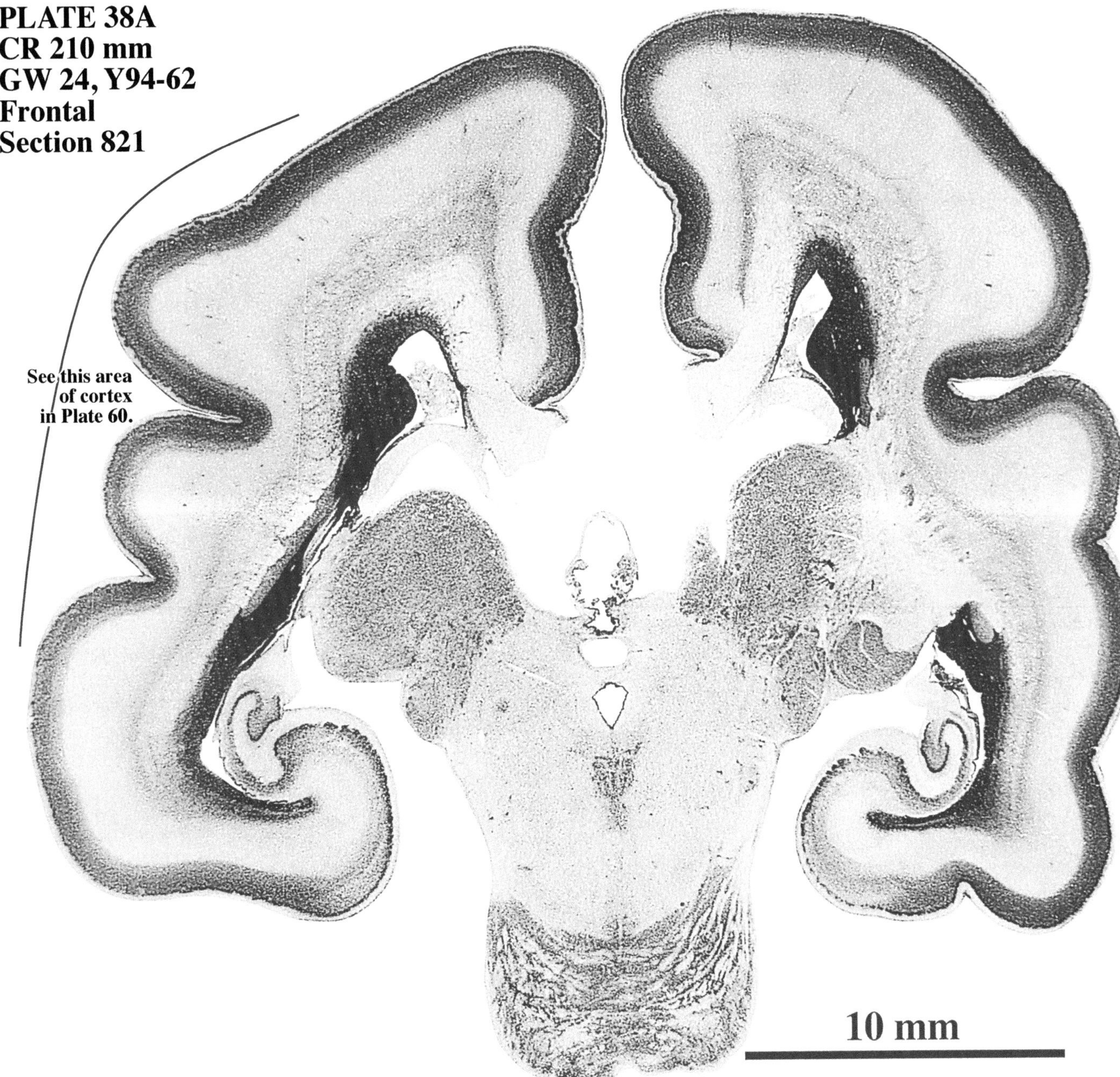

See detail of the brain core in Plates 51A and B.

LAYERS OF THE CORTICAL *STRATIFIED TRANSITIONAL FIELD (STF)*

STF1*—**Superficial fibrous layer with an early developmental stage ***(t1) when many cells are migrating through it, followed by a late stage ***(t2)*** with sparse cells. Endures as the subcortical white matter.

***STF2*—**Upper cellular layer, the last sojourn zone before cells translocate to the cortical plate.

STF3*—**Honeycomb trilaminar matrix ***(3a, 3b, 3c) of cells and fibers found only in granular cortices.

***STF4*—**Complex middle layer with three developmental stages:
***t1*–** fibrous layer without interspersed cells;
***t2*–** cells and fibers intermingle to form striations;
***t3*–** fibers endure in the deep white matter.

***STF5*—**Deep cellular layer, the first sojourn zone to appear outside the germinal matrix.

***STF6*—**Late-forming deep layer of callosal fibers outside the germinal matrix.

PLATE 38B

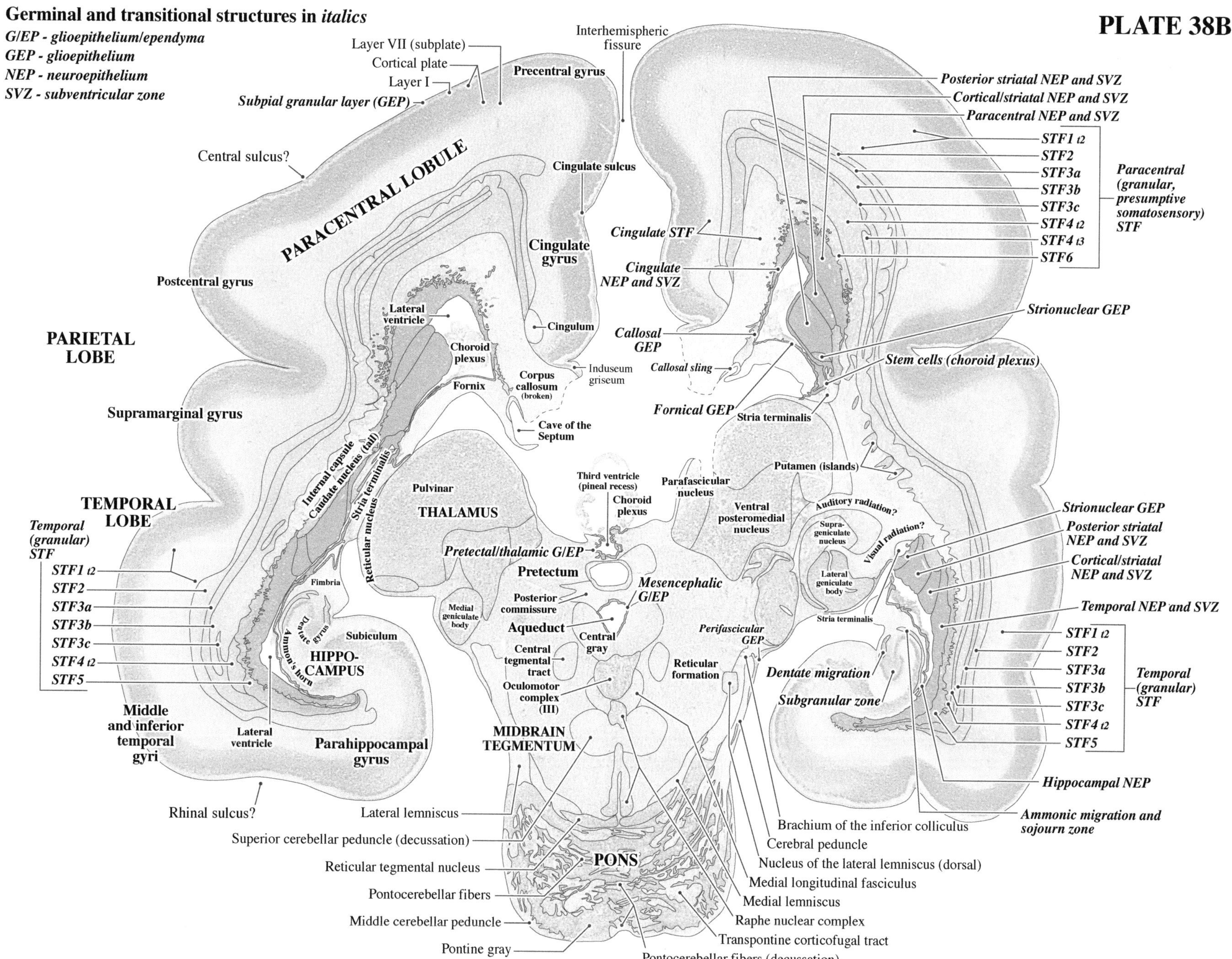

Germinal and transitional structures in *italics*
G/EP - glioepithelium/ependyma
GEP - glioepithelium
NEP - neuroepithelium
SVZ - subventricular zone
Layer VII (subplate)
Cortical plate
Layer I
Subpial granular layer (GEP)
Interhemispheric fissure
Precentral gyrus
Central sulcus?
PARACENTRAL LOBULE
Cingulate sulcus
Cingulate gyrus
Postcentral gyrus
PARIETAL LOBE
Supramarginal gyrus
Lateral ventricle
Choroid plexus
Fornix
Cingulum
Corpus callosum (broken)
Induseum griseum
Cave of the Septum
Internal capsule
Caudate nucleus (tail)
Stria terminalis
Reticular nucleus
Pulvinar
THALAMUS
TEMPORAL LOBE
Temporal (granular) STF
STF1 t2
STF2
STF3a
STF3b
STF3c
STF4 t2
STF5
Fimbria
Dentate gyrus
Ammon's horn
Subiculum
HIPPO-CAMPUS
Middle and inferior temporal gyri
Lateral ventricle
Parahippocampal gyrus
Rhinal sulcus?
Third ventricle (pineal recess)
Choroid plexus
Pretectal/thalamic G/EP
Pretectum
Medial geniculate body
Posterior commissure
Mesencephalic G/EP
Aqueduct
Central gray
Central tegmental tract
Oculomotor complex (III)
MIDBRAIN TEGMENTUM
Lateral lemniscus
Superior cerebellar peduncle (decussation)
Reticular tegmental nucleus
PONS
Pontocerebellar fibers
Middle cerebellar peduncle
Pontine gray
Posterior striatal NEP and SVZ
Cortical/striatal NEP and SVZ
Paracentral NEP and SVZ
STF1 t2
STF2
STF3a
STF3b
STF3c
STF4 t2
STF4 t3
STF6
Paracentral (granular, presumptive somatosensory) STF
Cingulate STF
Cingulate NEP and SVZ
Strionuclear GEP
Callosal GEP
Callosal sling
Stem cells (choroid plexus)
Fornical GEP
Stria terminalis
Putamen (islands)
Parafascicular nucleus
Ventral posteromedial nucleus
Auditory radiation?
Supra-geniculate nucleus
Visual radiation?
Lateral geniculate body
Stria terminalis
Strionuclear GEP
Posterior striatal NEP and SVZ
Cortical/striatal NEP and SVZ
Temporal NEP and SVZ
Perifascicular GEP
Reticular formation
Dentate migration
Subgranular zone
STF1 t2
STF2
STF3a
STF3b
STF3c
STF4 t2
STF5
Temporal (granular) STF
Hippocampal NEP
Ammonic migration and sojourn zone
Brachium of the inferior colliculus
Cerebral peduncle
Nucleus of the lateral lemniscus (dorsal)
Medial longitudinal fasciculus
Medial lemniscus
Raphe nuclear complex
Transpontine corticofugal tract
Pontocerebellar fibers (decussation)

PLATE 39A
CR 210 mm
GW 24, Y94-62
Frontal
Section 861

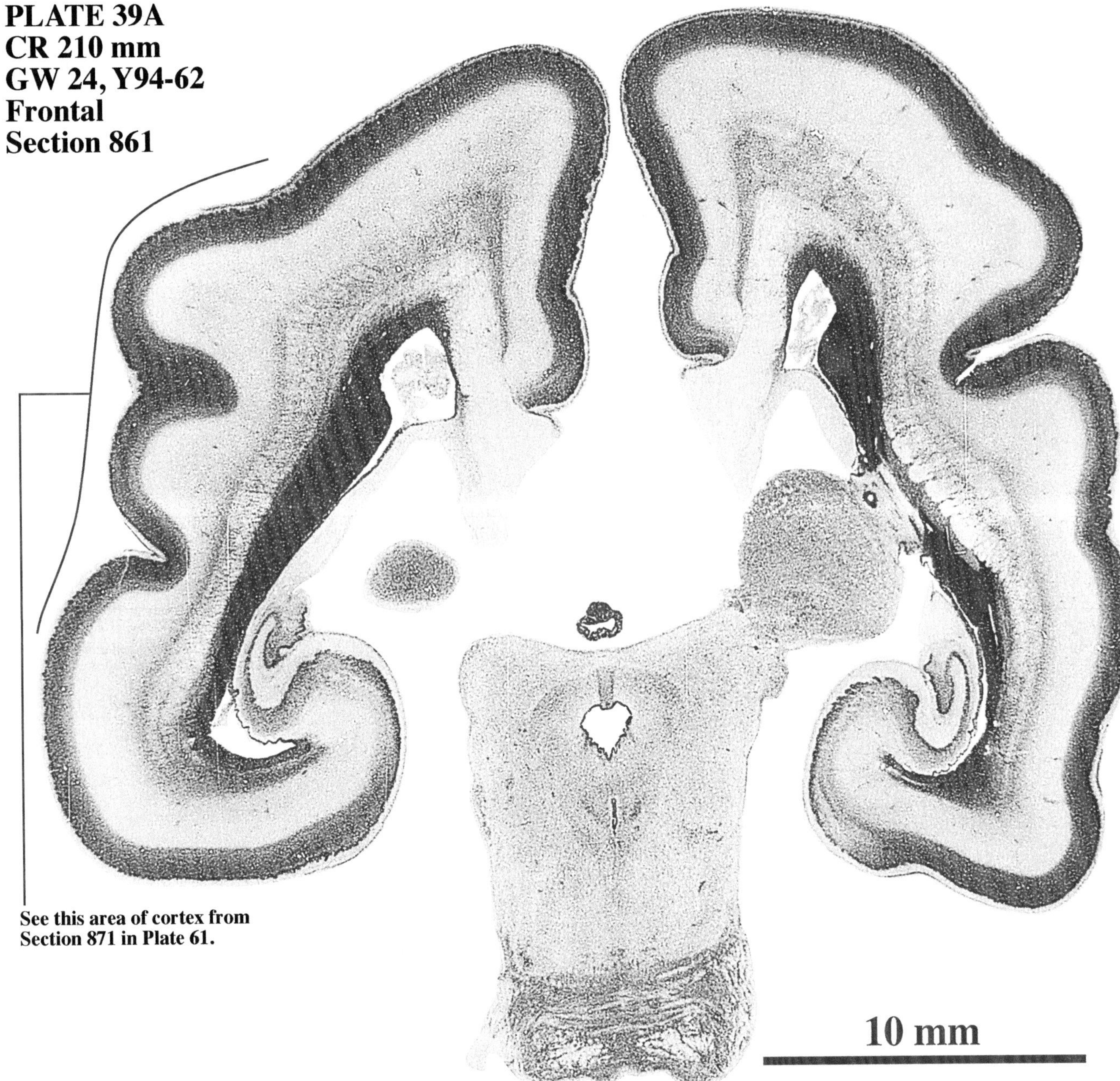

See this area of cortex from Section 871 in Plate 61.

See detail of brain core in Plates 52A and B.

LAYERS OF THE CORTICAL *STRATIFIED TRANSITIONAL FIELD (STF)*

STF1—Superficial fibrous layer with an early developmental stage ***(t1)*** when many cells are migrating through it, followed by a late stage ***(t2)*** with sparse cells. Endures as the subcortical white matter.

STF2—Upper cellular layer, the last sojourn zone before cells translocate to the cortical plate.

STF3—Honeycomb trilaminar matrix ***(3a, 3b, 3c)*** of cells and fibers found only in granular cortices.

STF4—Complex middle layer with three developmental stages:
t1– fibrous layer without interspersed cells;
t2– cells and fibers intermingle to form striations;
t3– fibers endure in the deep white matter.

STF5—Deep cellular layer, the first sojourn zone to appear outside the germinal matrix.

STF6—Late-forming deep layer of callosal fibers outside the germinal matrix.

PLATE 39B

Germinal and transitional structures in *italics*

G/EP - glioepithelium/ependyma
GEP - glioepithelium
NEP - neuroepithelium
SVZ - subventricular zone

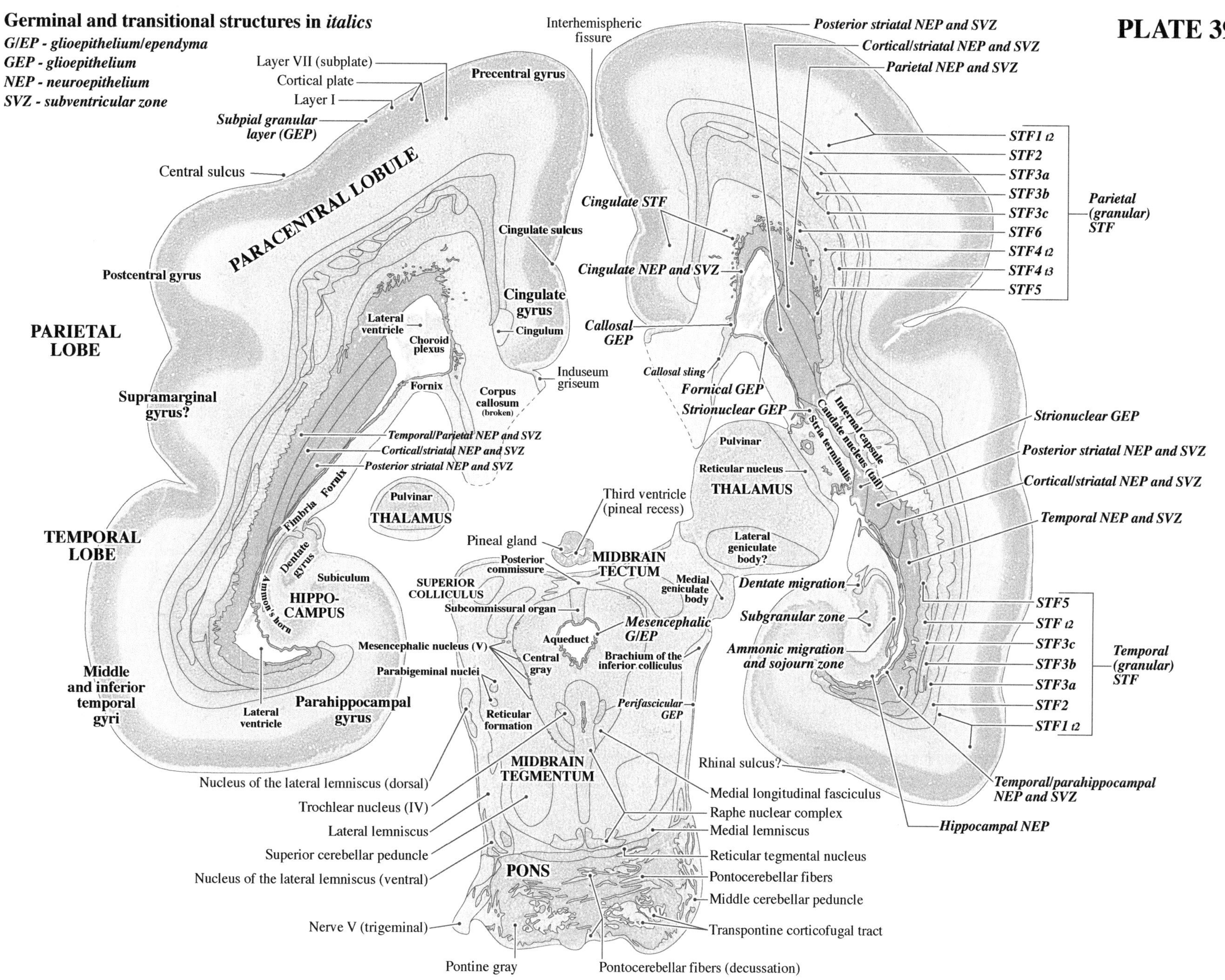

PLATE 40A
CR 210 mm
GW 24, Y94-62
Frontal
Section 901

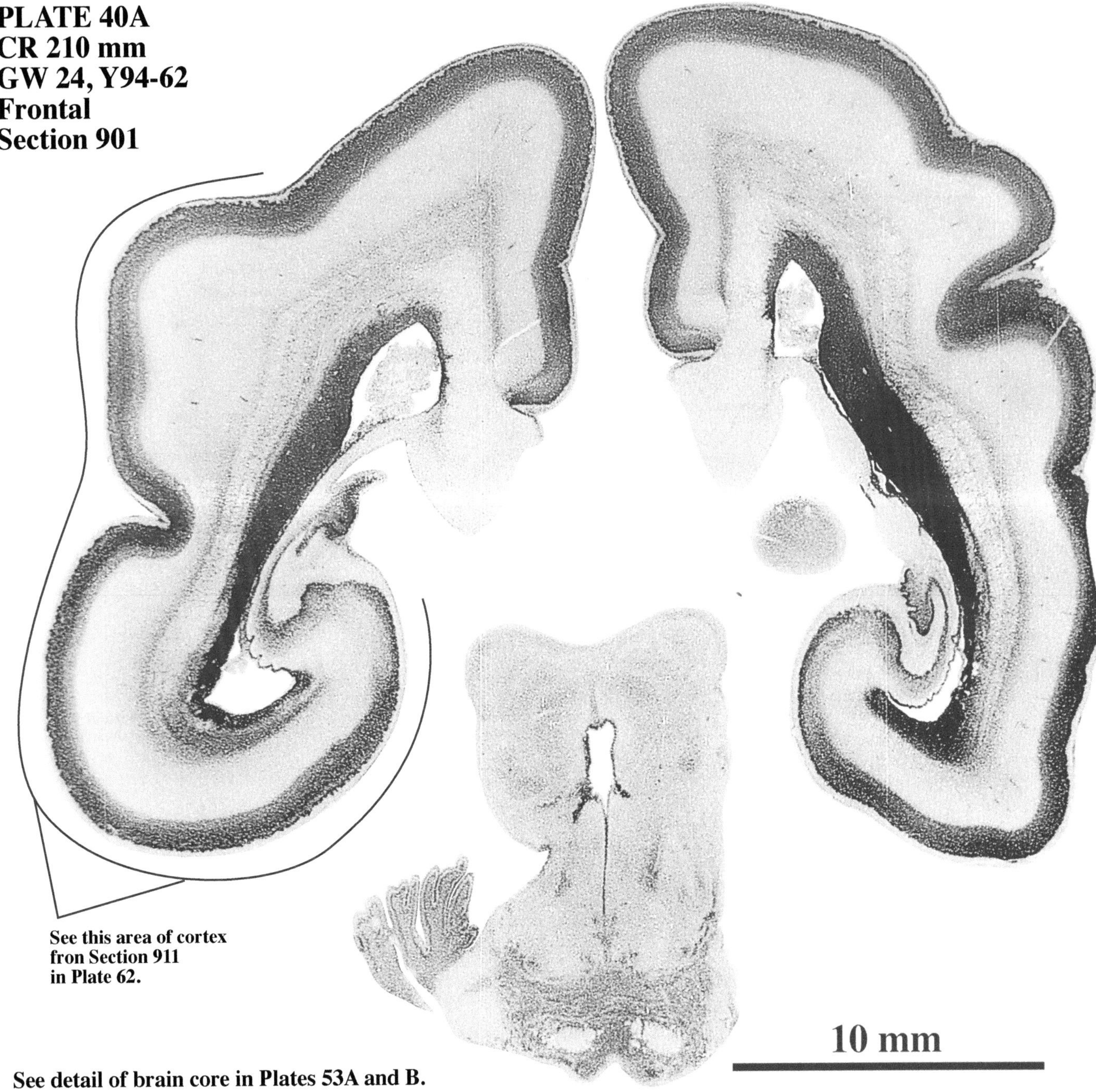

See this area of cortex fron Section 911 in Plate 62.

See detail of brain core in Plates 53A and B.

LAYERS OF THE CORTICAL *STRATIFIED TRANSITIONAL FIELD (STF)*

STF1*—**Superficial fibrous layer with an early developmental stage ***(t1) when many cells are migrating through it, followed by a late stage ***(t2)*** with sparse cells. Endures as the subcortical white matter.

***STF2*—**Upper cellular layer, the last sojourn zone before cells translocate to the cortical plate.

STF3*—**Honeycomb trilaminar matrix ***(3a, 3b, 3c) of cells and fibers found only in granular cortices.

***STF4*—**Complex middle layer with three developmental stages:
t1– fibrous layer without interspersed cells;
t2– cells and fibers intermingle to form striations;
t3– fibers endure in the deep white matter.

***STF5*—**Deep cellular layer, the first sojourn zone to appear outside the germinal matrix.

***STF6*—**Late-forming deep layer of callosal fibers outside the germinal matrix.

PLATE 40B

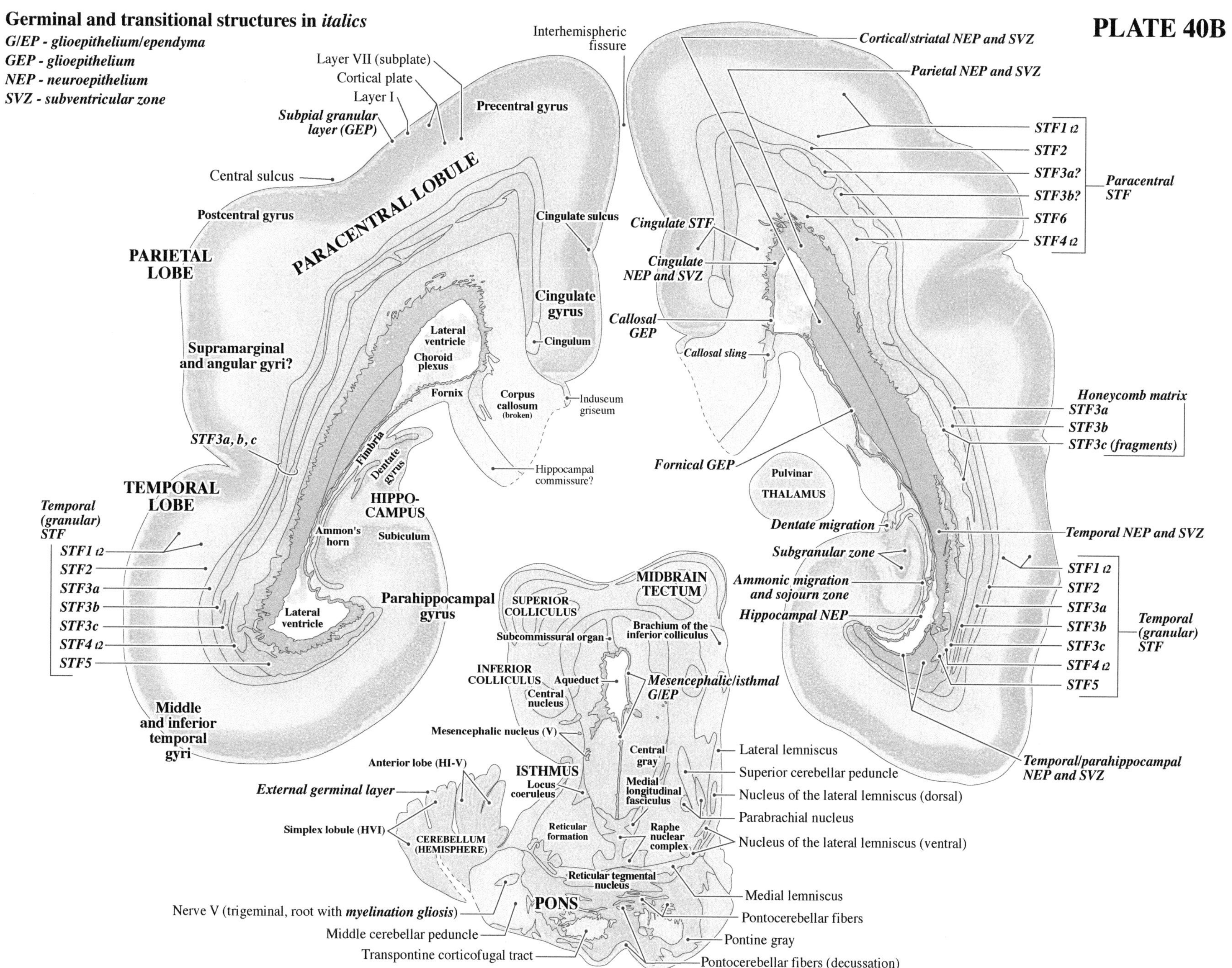
Germinal and transitional structures in *italics*
G/EP - glioepithelium/ependyma
GEP - glioepithelium
NEP - neuroepithelium
SVZ - subventricular zone
Interhemispheric fissure
Layer VII (subplate)
Cortical plate
Layer I
Subpial granular layer (GEP)
Precentral gyrus
Central sulcus
Postcentral gyrus
PARACENTRAL LOBULE
PARIETAL LOBE
Cingulate sulcus
Cingulate gyrus
Cingulum
Lateral ventricle
Choroid plexus
Fornix
Corpus callosum (broken)
Induseum griseum
Supramarginal and angular gyri?
STF3a, b, c
Fimbria
Dentate gyrus
Hippocampal commissure?
TEMPORAL LOBE
HIPPO-CAMPUS
Ammon's horn
Subiculum
Temporal (granular) STF
STF1 t2
STF2
STF3a
STF3b
STF3c
STF4 t2
STF5
Lateral ventricle
Parahippocampal gyrus
Middle and inferior temporal gyri
Cortical/striatal NEP and SVZ
Parietal NEP and SVZ
STF1 t2
STF2
STF3a?
STF3b?
STF6
STF4 t2
Paracentral STF
Cingulate STF
Cingulate NEP and SVZ
Callosal GEP
Callosal sling
Honeycomb matrix
STF3a
STF3b
STF3c (fragments)
Fornical GEP
Pulvinar
THALAMUS
Dentate migration
Temporal NEP and SVZ
Subgranular zone
Ammonic migration and sojourn zone
Hippocampal NEP
STF1 t2
STF2
STF3a
STF3b
STF3c
STF4 t2
STF5
Temporal (granular) STF
Temporal/parahippocampal NEP and SVZ
MIDBRAIN TECTUM
SUPERIOR COLLICULUS
Brachium of the inferior colliculus
Subcommissural organ
INFERIOR COLLICULUS
Aqueduct
Central nucleus
Mesencephalic/isthmal G/EP
Mesencephalic nucleus (V)
Central gray
Lateral lemniscus
Anterior lobe (HI-V)
ISTHMUS
Locus coeruleus
Medial longitudinal fasciculus
Superior cerebellar peduncle
External germinal layer
Nucleus of the lateral lemniscus (dorsal)
Parabrachial nucleus
Simplex lobule (HVI)
CEREBELLUM (HEMISPHERE)
Reticular formation
Raphe nuclear complex
Nucleus of the lateral lemniscus (ventral)
Reticular tegmental nucleus
Medial lemniscus
PONS
Nerve V (trigeminal, root with *myelination gliosis*)
Pontocerebellar fibers
Middle cerebellar peduncle
Pontine gray
Transpontine corticofugal tract
Pontocerebellar fibers (decussation)

PLATE 41A
CR 210 mm
GW 24, Y94-62
Frontal
Section 941

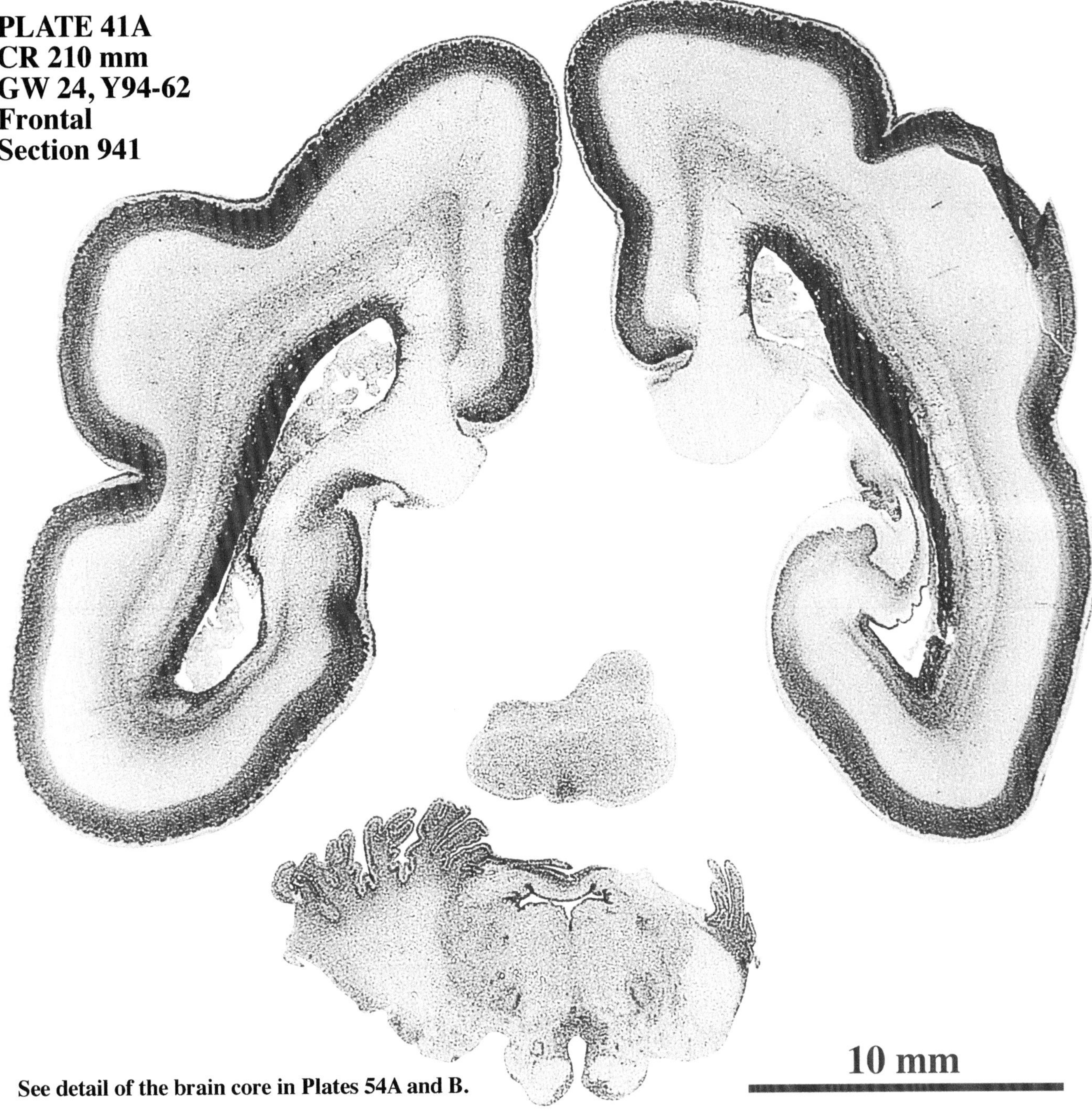

See detail of the brain core in Plates 54A and B.

LAYERS OF THE CORTICAL *STRATIFIED TRANSITIONAL FIELD (STF)*

STF1*—**Superficial fibrous layer with an early developmental stage ***(t1) when many cells are migrating through it, followed by a late stage ***(t2)*** with sparse cells. Endures as the subcortical white matter.

***STF2*—**Upper cellular layer, the last sojourn zone before cells translocate to the cortical plate.

STF3*—**Honeycomb trilaminar matrix ***(3a, 3b, 3c) of cells and fibers found only in granular cortices.

***STF4*—**Complex middle layer with three developmental stages:
t1– fibrous layer without interspersed cells;
t2– cells and fibers intermingle to form striations;
t3– fibers endure in the deep white matter.

***STF5*—**Deep cellular layer, the first sojourn zone to appear outside the germinal matrix.

***STF6*—**Late-forming deep layer of callosal fibers outside the germinal matrix.

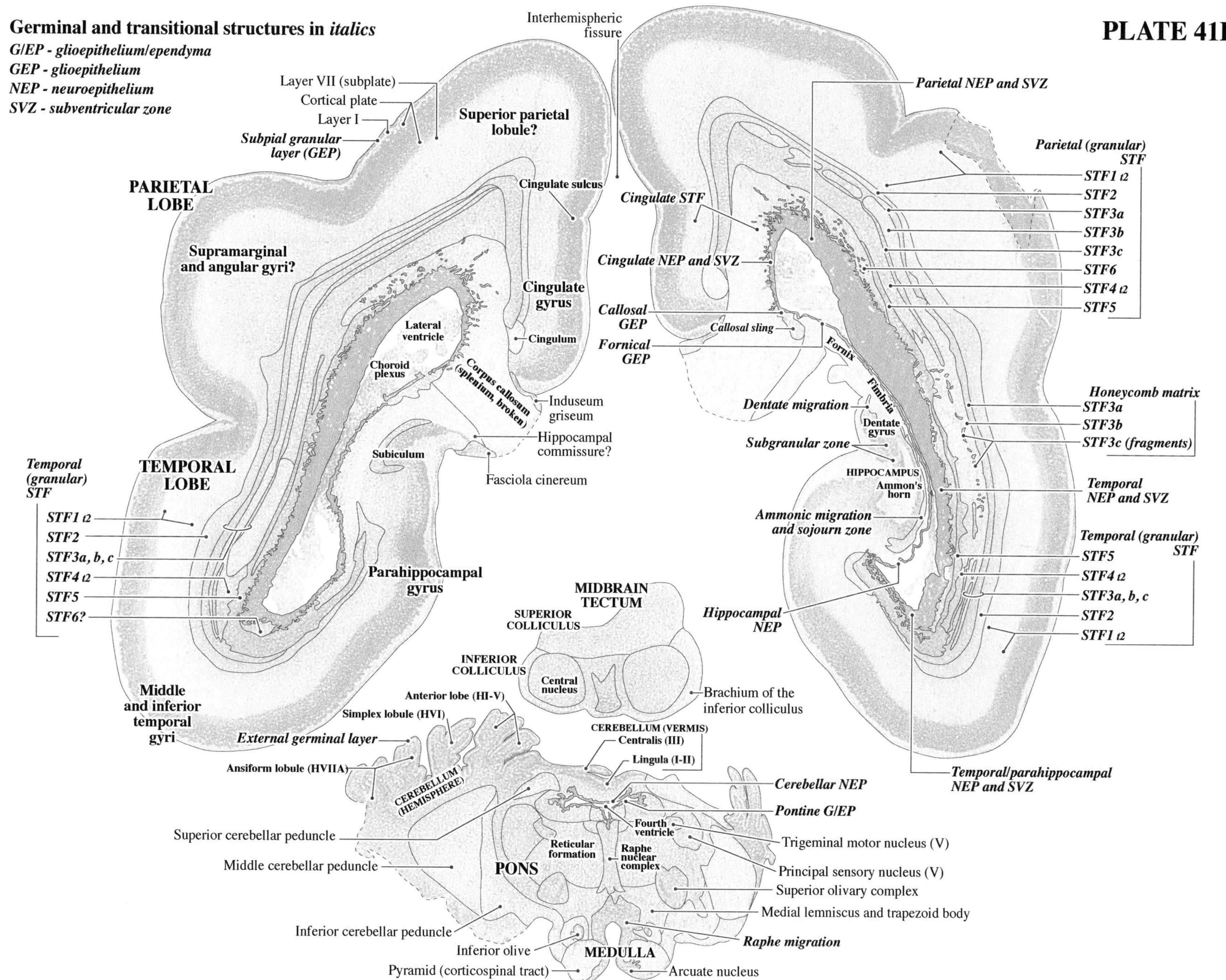

PLATE 41B
Germinal and transitional structures in italics
G/EP - glioepithelium/ependyma
GEP - glioepithelium
NEP - neuroepithelium
SVZ - subventricular zone
Interhemispheric fissure
Layer VII (subplate)
Cortical plate
Layer I
Subpial granular layer (GEP)
Superior parietal lobule?
PARIETAL LOBE
Cingulate sulcus
Supramarginal and angular gyri?
Cingulate gyrus
Lateral ventricle
Cingulum
Choroid plexus
Corpus callosum (splenium, broken)
Induseum griseum
Hippocampal commissure?
Subiculum
Fasciola cinereum
TEMPORAL LOBE
Temporal (granular) STF
STF1 t2
STF2
STF3a, b, c
STF4 t2
STF5
STF6?
Parahippocampal gyrus
Middle and inferior temporal gyri
Parietal NEP and SVZ
Parietal (granular) STF
STF1 t2
STF2
STF3a
STF3b
STF3c
STF6
STF4 t2
STF5
Cingulate STF
Cingulate NEP and SVZ
Callosal GEP
Callosal sling
Fornical GEP
Fornix
Fimbria
Dentate migration
Dentate gyrus
Subgranular zone
HIPPOCAMPUS
Ammon's horn
Honeycomb matrix
STF3a
STF3b
STF3c (fragments)
Temporal NEP and SVZ
Ammonic migration and sojourn zone
Temporal (granular) STF
STF5
STF4 t2
STF3a, b, c
STF2
STF1 t2
Hippocampal NEP
Temporal/parahippocampal NEP and SVZ
MIDBRAIN TECTUM
SUPERIOR COLLICULUS
INFERIOR COLLICULUS
Central nucleus
Brachium of the inferior colliculus
Anterior lobe (HI-V)
Simplex lobule (HVI)
External germinal layer
Ansiform lobule (HVIIA)
CEREBELLUM (HEMISPHERE)
CEREBELLUM (VERMIS)
Centralis (III)
Lingula (I-II)
Cerebellar NEP
Pontine G/EP
Fourth ventricle
Trigeminal motor nucleus (V)
Principal sensory nucleus (V)
Superior cerebellar peduncle
Reticular formation
Raphe nuclear complex
Middle cerebellar peduncle
PONS
Superior olivary complex
Medial lemniscus and trapezoid body
Inferior cerebellar peduncle
Raphe migration
Inferior olive
MEDULLA
Pyramid (corticospinal tract)
Arcuate nucleus

PLATE 42A
CR 210 mm
GW 24, Y94-62
Frontal
Section 981

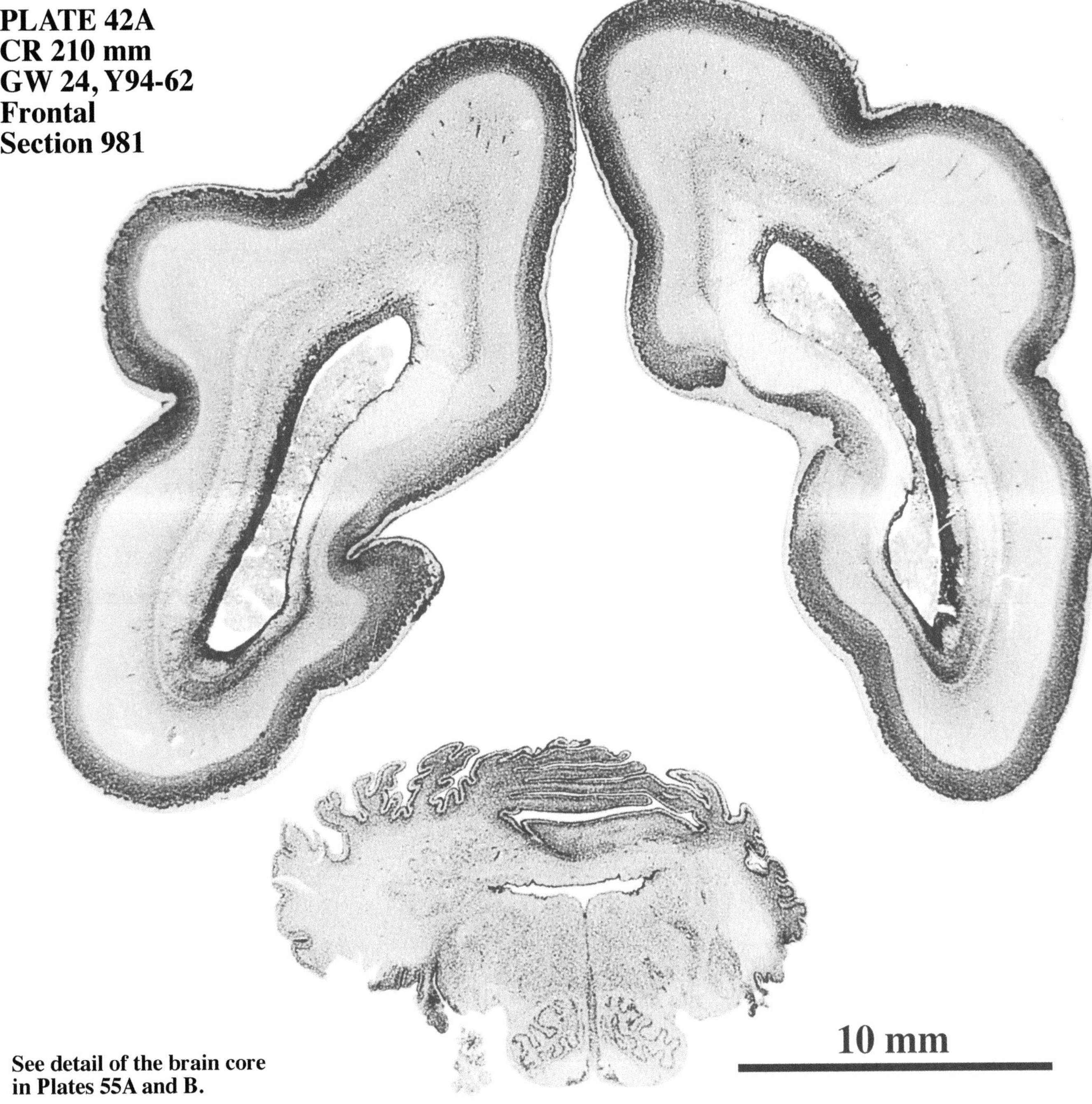

See detail of the brain core in Plates 55A and B.

LAYERS OF THE CORTICAL *STRATIFIED TRANSITIONAL FIELD (STF)*

STF1—Superficial fibrous layer with an early developmental stage ***(t1)*** when many cells are migrating through it, followed by a late stage ***(t2)*** with sparse cells. Endures as the subcortical white matter.

STF2—Upper cellular layer, the last sojourn zone before cells translocate to the cortical plate.

STF3—Honeycomb trilaminar matrix ***(3a, 3b, 3c)*** of cells and fibers found only in granular cortices.

STF4—Complex middle layer with three developmental stages:
t1– fibrous layer without interspersed cells;
t2– cells and fibers intermingle to form striations;
t3– fibers endure in the deep white matter.

STF5—Deep cellular layer, the first sojourn zone to appear outside the germinal matrix.

STF6—Late-forming deep layer of callosal fibers outside the germinal matrix.

PLATE 42B

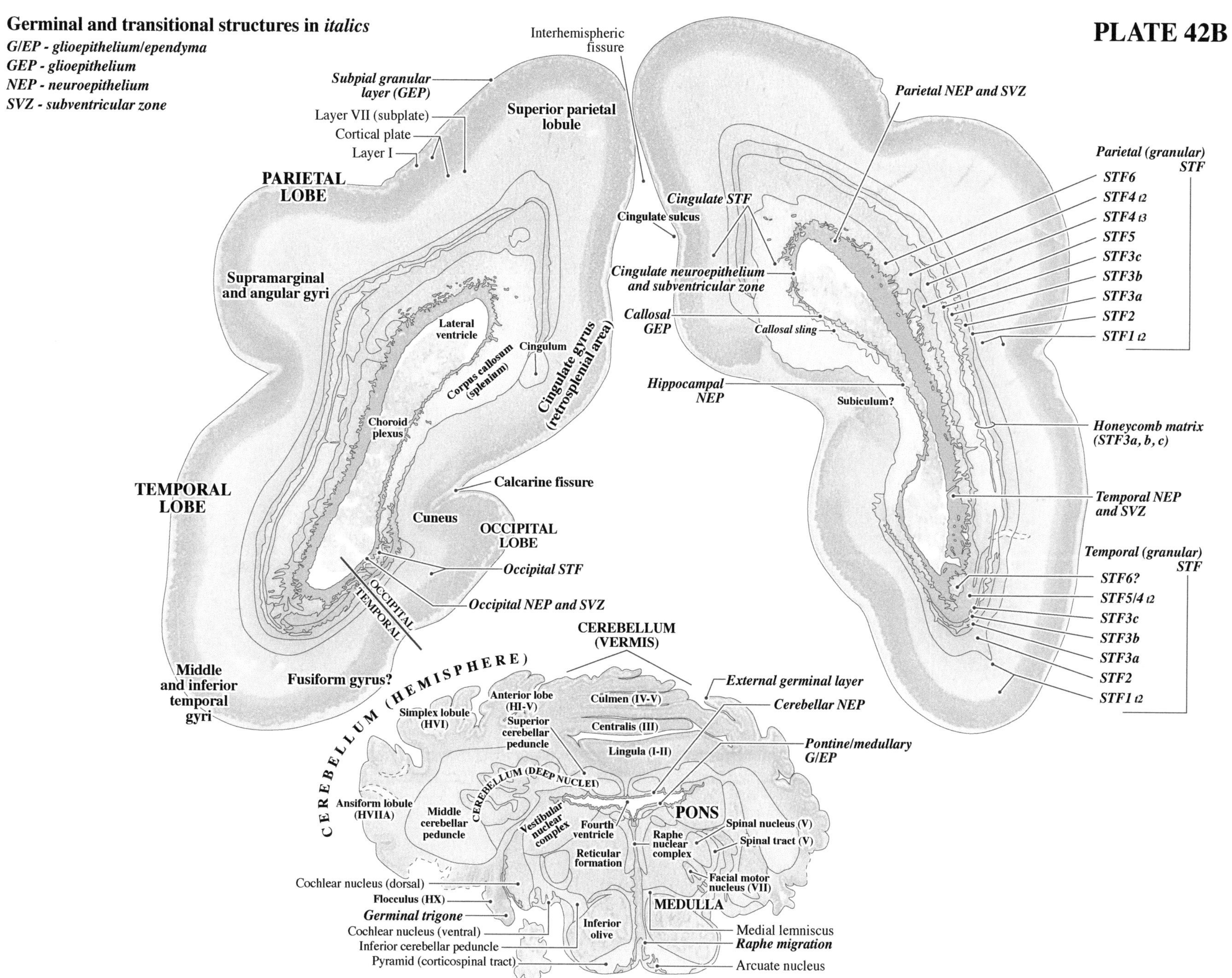

Germinal and transitional structures in *italics*
G/EP - glioepithelium/ependyma
GEP - glioepithelium
NEP - neuroepithelium
SVZ - subventricular zone
Interhemispheric fissure
Subpial granular layer (GEP)
Layer VII (subplate)
Cortical plate
Layer I
Superior parietal lobule
PARIETAL LOBE
Supramarginal and angular gyri
Lateral ventricle
Cingulum
Corpus callosum (splenium)
Cingulate gyrus (retrosplenial area)
Choroid plexus
Calcarine fissure
TEMPORAL LOBE
Cuneus
OCCIPITAL LOBE
Occipital STF
Occipital NEP and SVZ
OCCIPITAL
TEMPORAL
Middle and inferior temporal gyri
Fusiform gyrus?
Cingulate sulcus
Cingulate STF
Cingulate neuroepithelium and subventricular zone
Callosal GEP
Callosal sling
Hippocampal NEP
Subiculum?
Parietal NEP and SVZ
Parietal (granular) STF
STF6
STF4 t2
STF4 t3
STF5
STF3c
STF3b
STF3a
STF2
STF1 t2
Honeycomb matrix (STF3a, b, c)
Temporal NEP and SVZ
Temporal (granular) STF
STF6?
STF5/4 t2
STF3c
STF3b
STF3a
STF2
STF1 t2
CEREBELLUM (VERMIS)
CEREBELLUM (HEMISPHERE)
External germinal layer
Cerebellar NEP
Anterior lobe (HI-V)
Simplex lobule (HVI)
Culmen (IV-V)
Superior cerebellar peduncle
Centralis (III)
Lingula (I-II)
Pontine/medullary G/EP
CEREBELLUM (DEEP NUCLEI)
Ansiform lobule (HVIIA)
Middle cerebellar peduncle
Vestibular nuclear complex
Fourth ventricle
PONS
Spinal nucleus (V)
Spinal tract (V)
Raphe nuclear complex
Reticular formation
Facial motor nucleus (VII)
Cochlear nucleus (dorsal)
Flocculus (HX)
Germinal trigone
Cochlear nucleus (ventral)
Inferior cerebellar peduncle
Pyramid (corticospinal tract)
Inferior olive
MEDULLA
Medial lemniscus
Raphe migration
Arcuate nucleus

PLATE 43A
CR 210 mm
GW 24, Y94-62
Frontal
Section 1041

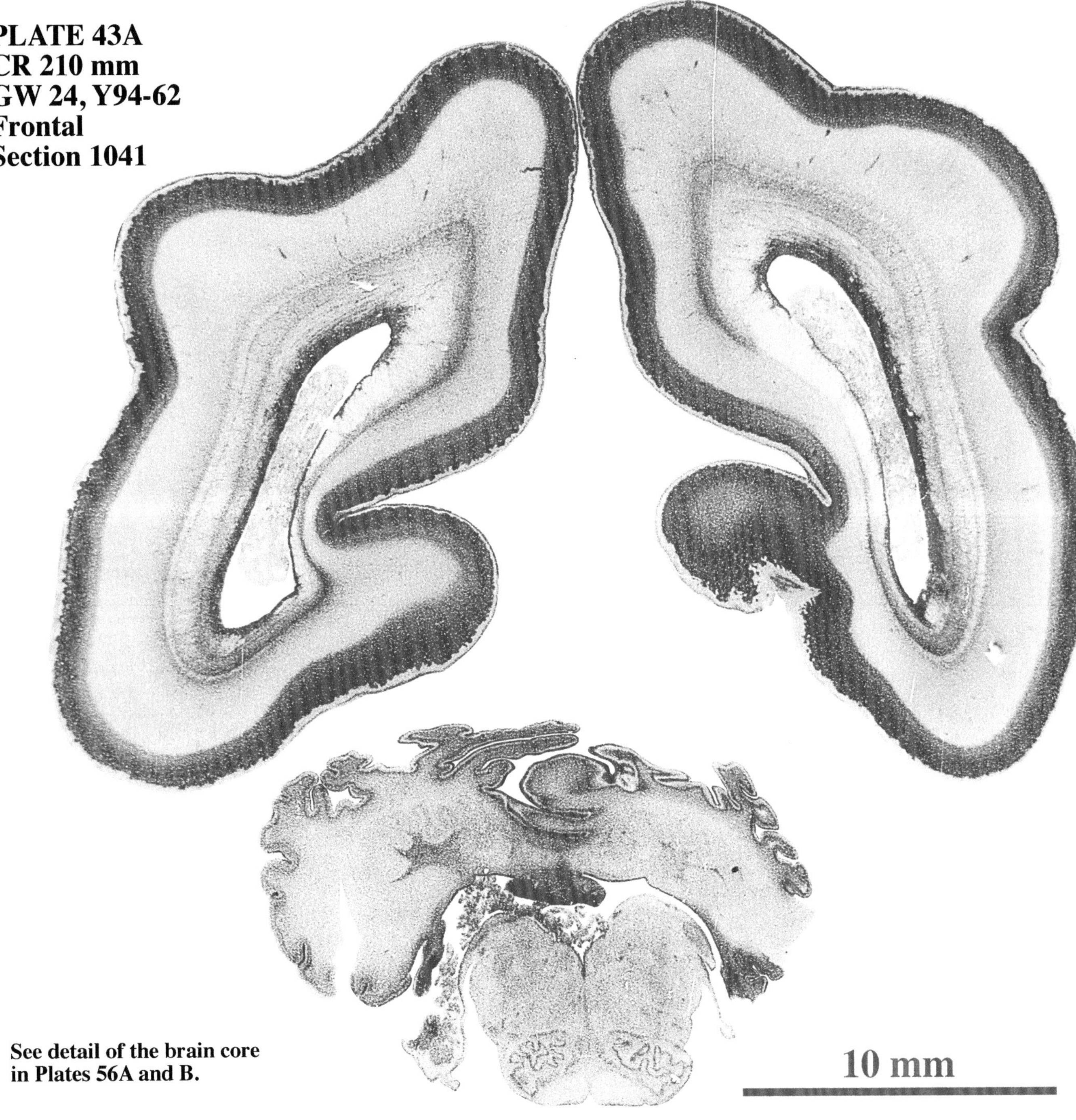

See detail of the brain core in Plates 56A and B.

LAYERS OF THE CORTICAL *STRATIFIED TRANSITIONAL FIELD (STF)*

STF1 — Superficial fibrous layer with an early developmental stage ***(t1)*** when many cells are migrating through it, followed by a late stage ***(t2)*** with sparse cells. Endures as the subcortical white matter.

STF2 — Upper cellular layer, the last sojourn zone before cells translocate to the cortical plate.

STF3 — Honeycomb trilaminar matrix ***(3a, 3b, 3c)*** of cells and fibers found only in granular cortices.

STF4 — Complex middle layer with three developmental stages:
t1 – fibrous layer without interspersed cells;
t2 – cells and fibers intermingle to form striations;
t3 – fibers endure in the deep white matter.

STF5 — Deep cellular layer, the first sojourn zone to appear outside the germinal matrix.

STF6 — Late-forming deep layer of callosal fibers outside the germinal matrix.

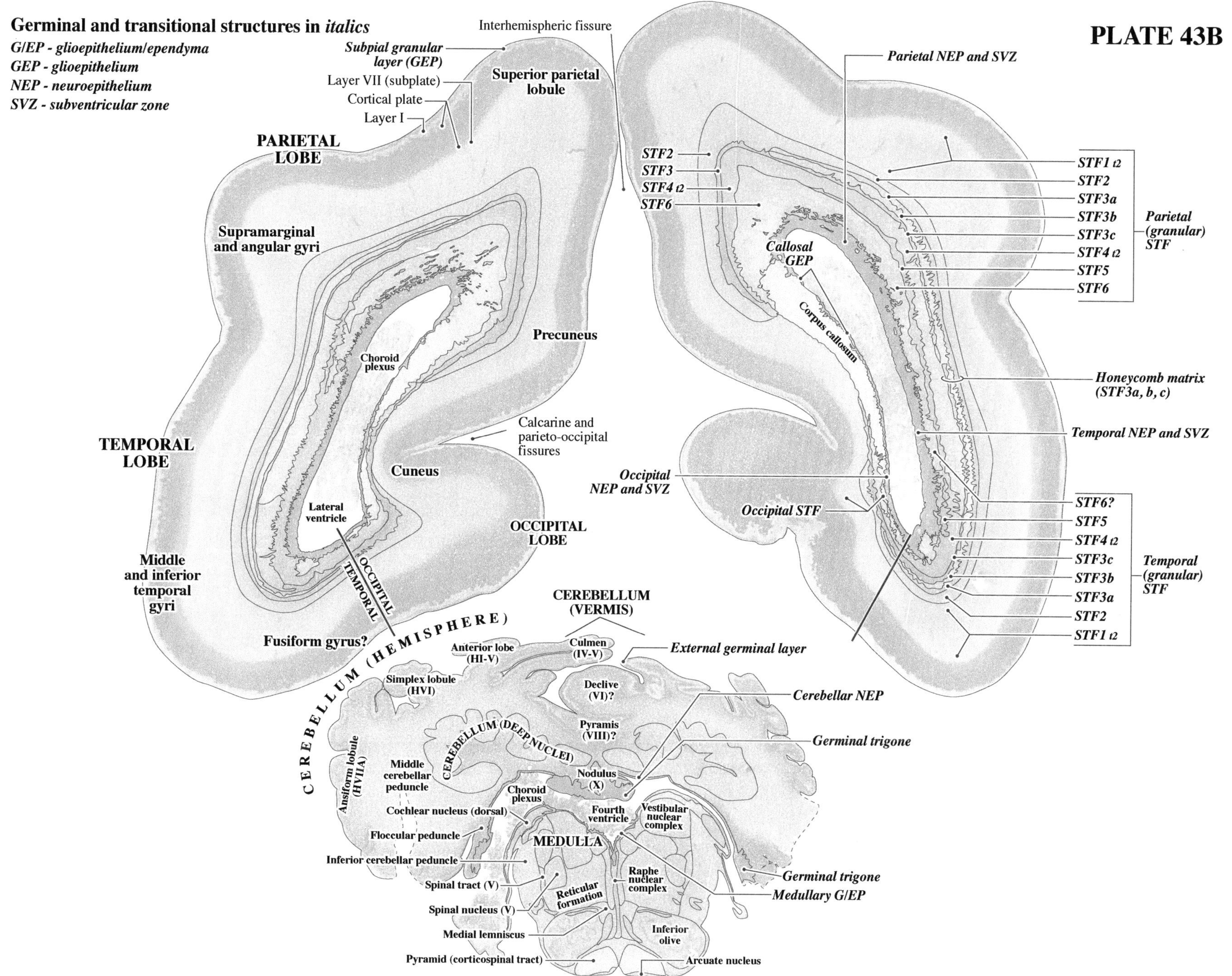
PLATE 43B
Germinal and transitional structures in *italics*
G/EP - glioepithelium/ependyma
GEP - glioepithelium
NEP - neuroepithelium
SVZ - subventricular zone
Interhemispheric fissure
Subpial granular layer (GEP)
Layer VII (subplate)
Cortical plate
Layer I
Superior parietal lobule
PARIETAL LOBE
Supramarginal and angular gyri
Choroid plexus
Precuneus
Calcarine and parieto-occipital fissures
Cuneus
TEMPORAL LOBE
Lateral ventricle
OCCIPITAL LOBE
OCCIPITAL
TEMPORAL
Middle and inferior temporal gyri
Fusiform gyrus?
Parietal NEP and SVZ
STF2
STF3
STF4 t2
STF6
Callosal GEP
Corpus callosum
STF1 t2
STF2
STF3a
STF3b
STF3c
STF4 t2
STF5
STF6
Parietal (granular) STF
Honeycomb matrix (STF3a, b, c)
Temporal NEP and SVZ
Occipital NEP and SVZ
Occipital STF
STF6?
STF5
STF4 t2
STF3c
STF3b
STF3a
STF2
STF1 t2
Temporal (granular) STF
CEREBELLUM (VERMIS)
CEREBELLUM (HEMISPHERE)
Anterior lobe (HI-V)
Culmen (IV-V)
External germinal layer
Simplex lobule (HVI)
Declive (VI)?
Cerebellar NEP
Pyramis (VIII)?
CEREBELLUM (DEEP NUCLEI)
Germinal trigone
Ansiform lobule (HVIIA)
Middle cerebellar peduncle
Nodulus (X)
Choroid plexus
Cochlear nucleus (dorsal)
Fourth ventricle
Vestibular nuclear complex
Floccular peduncle
MEDULLA
Inferior cerebellar peduncle
Spinal tract (V)
Reticular formation
Raphe nuclear complex
Germinal trigone
Medullary G/EP
Spinal nucleus (V)
Medial lemniscus
Inferior olive
Pyramid (corticospinal tract)
Arcuate nucleus

PLATE 44A
CR 210 mm
GW 24, Y94-62
Frontal
Section 1141

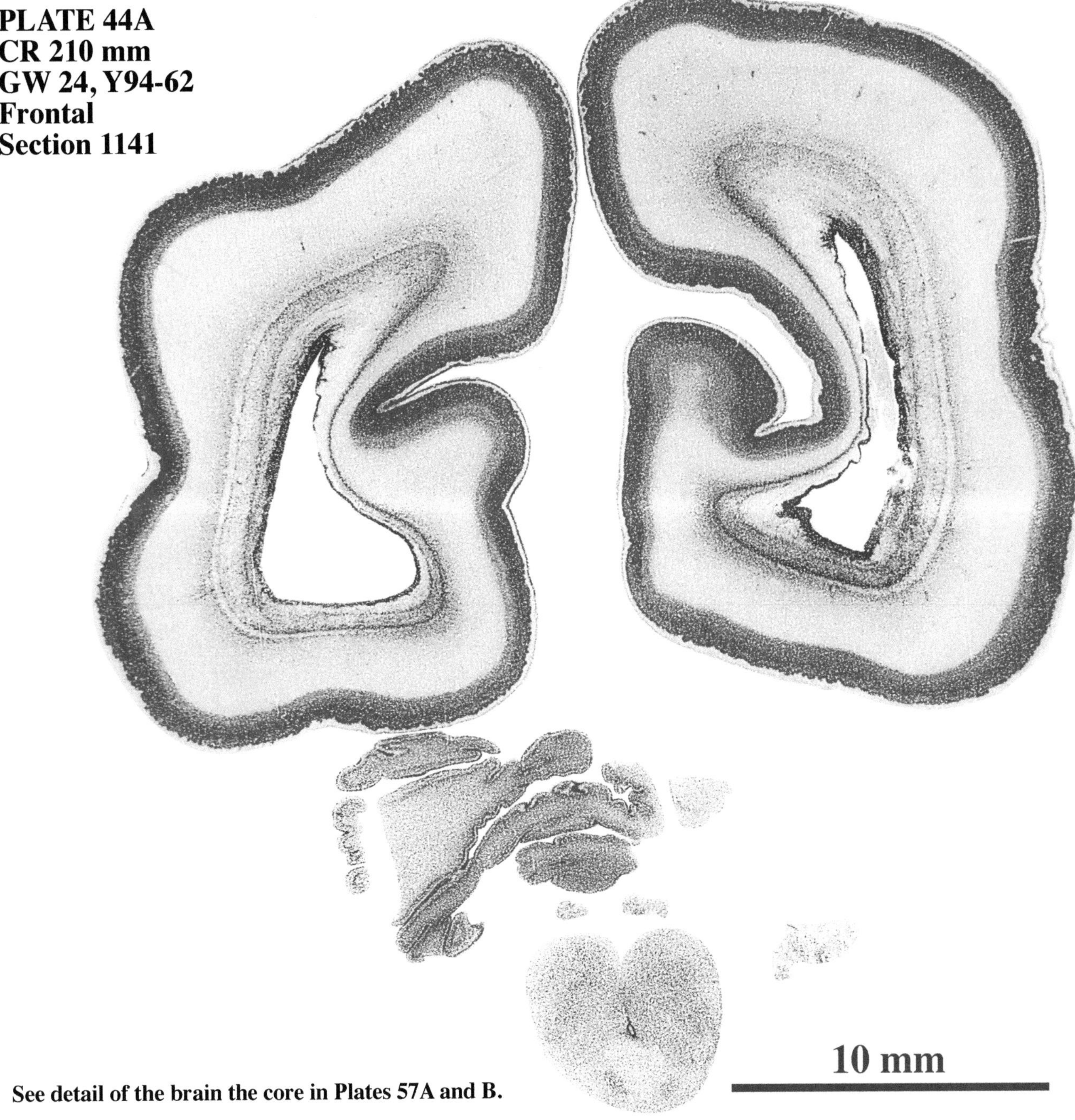

See detail of the brain the core in Plates 57A and B.

LAYERS OF THE CORTICAL *STRATIFIED TRANSITIONAL FIELD (STF)*

STF1—Superficial fibrous layer with an early developmental stage ***(t1)*** when many cells are migrating through it, followed by a late stage ***(t2)*** with sparse cells. Endures as the subcortical white matter.

STF2—Upper cellular layer, the last sojourn zone before cells translocate to the cortical plate.

STF3—Honeycomb trilaminar matrix ***(3a, 3b, 3c)*** of cells and fibers found only in granular cortices.

STF4—Complex middle layer with three developmental stages:
t1– fibrous layer without interspersed cells;
t2– cells and fibers intermingle to form striations;
t3– fibers endure in the deep white matter.

STF5—Deep cellular layer, the first sojourn zone to appear outside the germinal matrix.

STF6—Late-forming deep layer of callosal fibers outside the germinal matrix.

PLATE 44B

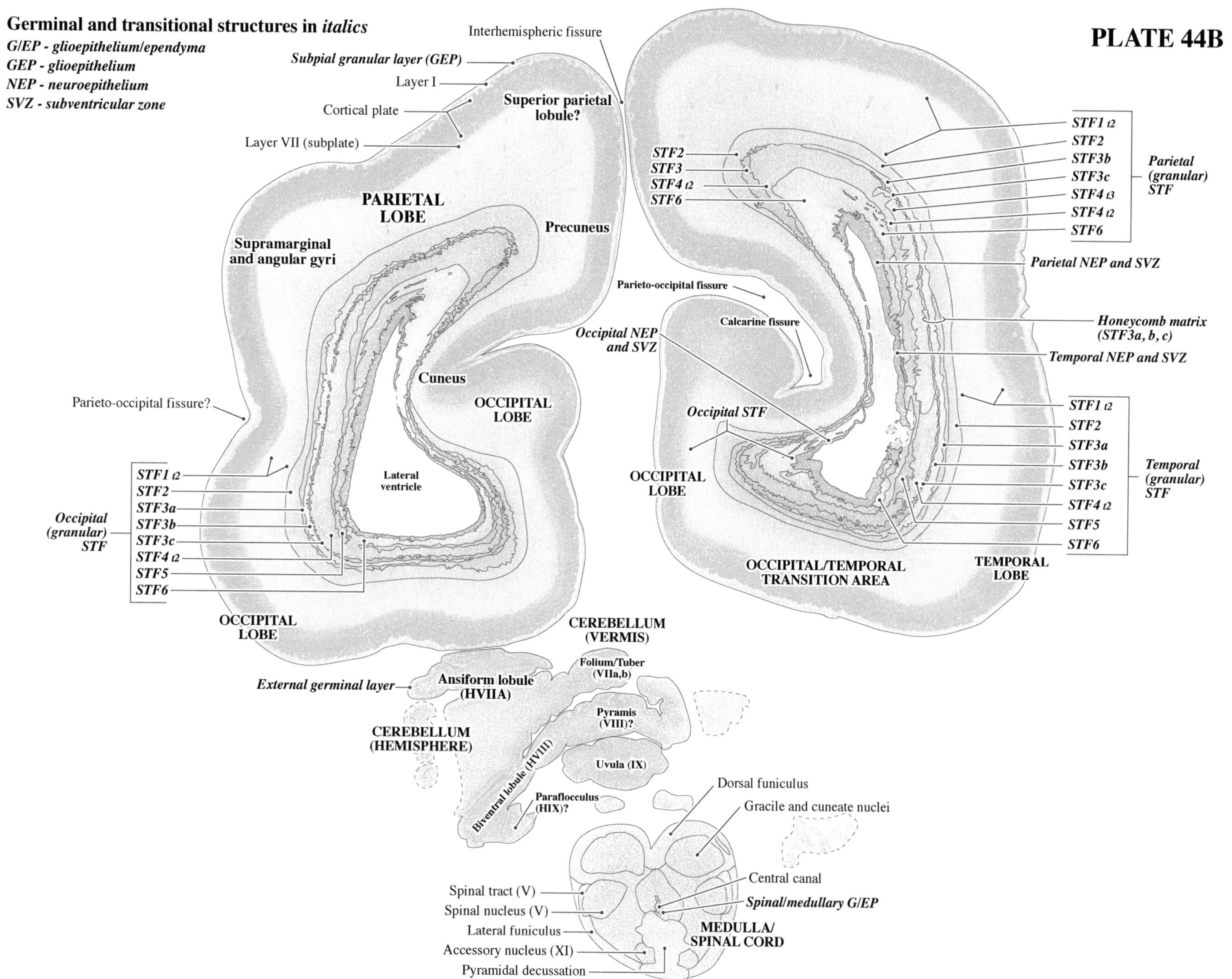

Germinal and transitional structures in *italics*
G/EP - glioepithelium/ependyma
GEP - glioepithelium
NEP - neuroepithelium
SVZ - subventricular zone
Interhemispheric fissure
Subpial granular layer (GEP)
Layer I
Cortical plate
Layer VII (subplate)
Superior parietal lobule?
PARIETAL LOBE
Precuneus
Supramarginal and angular gyri
Cuneus
OCCIPITAL LOBE
Parieto-occipital fissure?
Lateral ventricle
STF1 t2
STF2
STF3a
STF3b
STF3c
STF4 t2
STF5
STF6
Occipital (granular) STF
OCCIPITAL LOBE
STF2
STF3
STF4 t2
STF6
STF1 t2
STF2
STF3b
STF3c
STF4 t3
STF4 t2
STF6
Parietal (granular) STF
Parietal NEP and SVZ
Parieto-occipital fissure
Calcarine fissure
Occipital NEP and SVZ
Honeycomb matrix (STF3a, b, c)
Temporal NEP and SVZ
Occipital STF
OCCIPITAL LOBE
STF1 t2
STF2
STF3a
STF3b
STF3c
STF4 t2
STF5
STF6
Temporal (granular) STF
OCCIPITAL/TEMPORAL TRANSITION AREA
TEMPORAL LOBE
CEREBELLUM (VERMIS)
Folium/Tuber (VIIa,b)
External germinal layer
Ansiform lobule (HVIIA)
CEREBELLUM (HEMISPHERE)
Pyramis (VIII)?
Uvula (IX)
Biventral lobule (HVIII)
Paraflocculus (HIX)?
Dorsal funiculus
Gracile and cuneate nuclei
Central canal
Spinal/medullary G/EP
Spinal tract (V)
Spinal nucleus (V)
Lateral funiculus
Accessory nucleus (XI)
Pyramidal decussation
MEDULLA/ SPINAL CORD

PLATE 45A
CR 210 mm
GW 24, Y94-62
Frontal
Section 1201

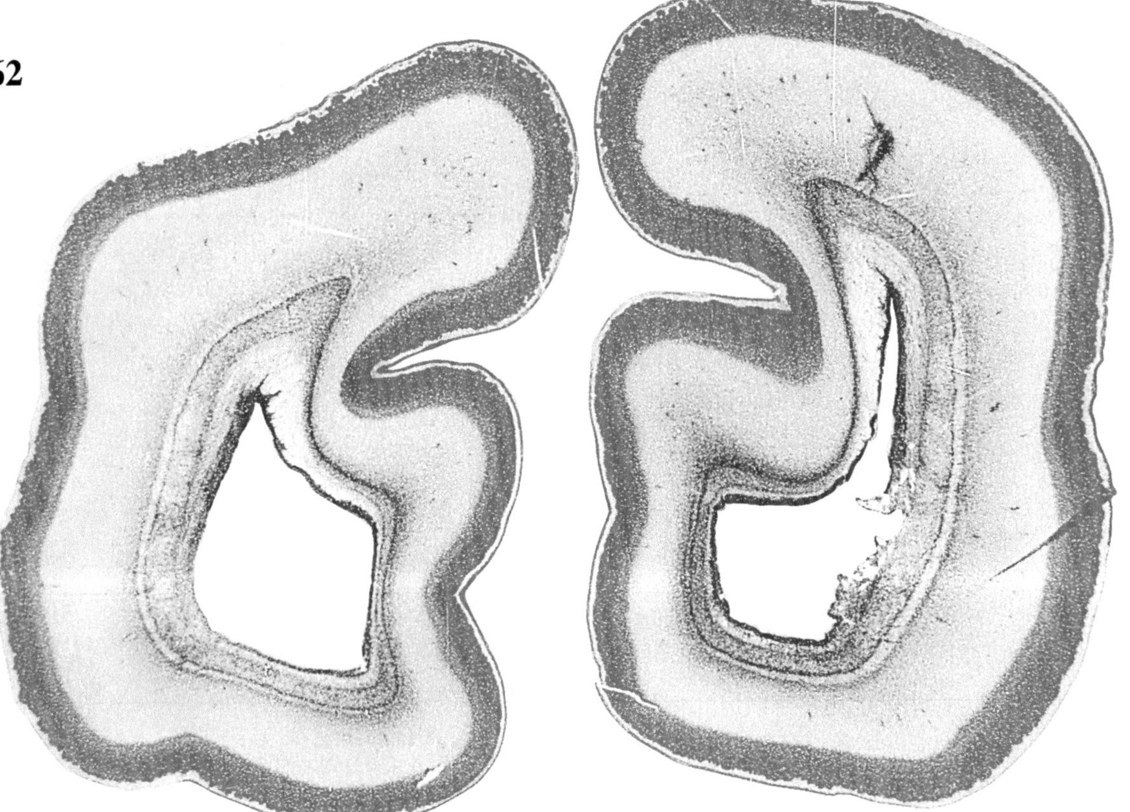

LAYERS OF THE CORTICAL *STRATIFIED TRANSITIONAL FIELD (STF)*

STF1—Superficial fibrous layer with an early developmental stage ***(t1)*** when many cells are migrating through it, followed by a late stage ***(t2)*** with sparse cells. Endures as the subcortical white matter.

STF2—Upper cellular layer, the last sojourn zone before cells translocate to the cortical plate.

STF3—Honeycomb trilaminar matrix ***(3a, 3b, 3c)*** of cells and fibers found only in granular cortices.

STF4—Complex middle layer with three developmental stages:
t1– fibrous layer without interspersed cells;
t2– cells and fibers intermingle to form striations;
t3– fibers endure in the deep white matter.

STF5—Deep cellular layer, the first sojourn zone to appear outside the germinal matrix.

STF6—Late-forming deep layer of callosal fibers outside the germinal matrix.

10 mm

See detail of the brain core in Plates 58A and B.

PLATE 45B

Germinal and transitional structures in *italics*

G/EP - glioepithelium/ependyma
GEP - glioepithelium
NEP - neuroepithelium
SVZ - subventricular zone

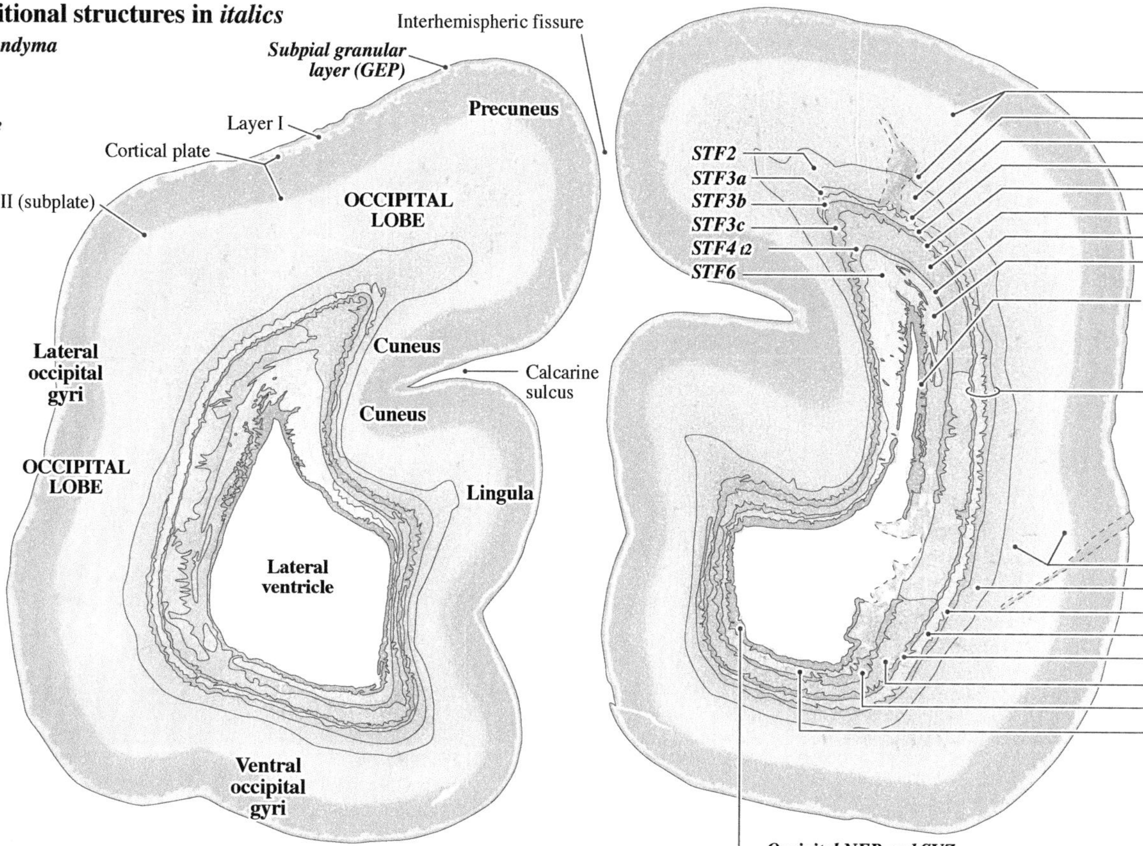

Interhemispheric fissure
Subpial granular layer (GEP)
Layer I
Cortical plate
Layer VII (subplate)
Precuneus
OCCIPITAL LOBE
Lateral occipital gyri
OCCIPITAL LOBE
Cuneus
Cuneus
Calcarine sulcus
Lingula
Lateral ventricle
Ventral occipital gyri
STF2
STF3a
STF3b
STF3c
STF4 t2
STF6
STF t2
STF2
STF3a
STF3b
STF3c
STF4 t2
STF5
STF6
Occipital (granular) STF
Occipital NEP and SVZ
Honeycomb matrix (STF3a, b, c)
STF1 t2
STF2
STF3a
STF3b
STF3c
STF4 t2
STF5
STF6
Occipital (granular) STF
Occipital NEP and SVZ

SPINAL CORD

Dorsal funiculus
Central canal surrounded by ***spinal G/EP***
Dorsal gray matter
Subpial granular layer (GEP surrounding fiber tracts)
Spinal nucleus (V)?
Spinal tract (V)?
Central autonomic area
Lateral corticospinal tract (crossed fibers)
Ventral gray matter
Lateral funiculus
Ventral funiculus

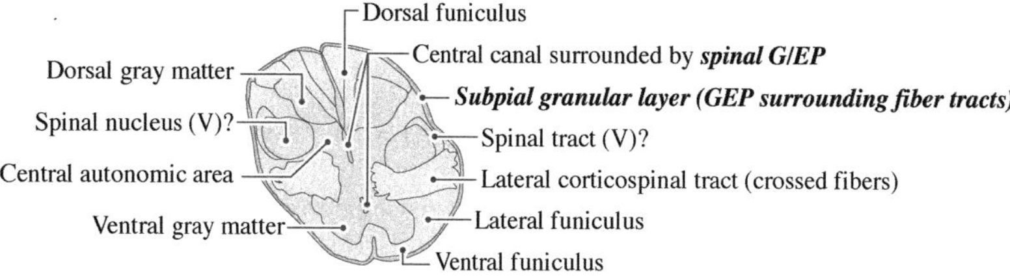

PLATE 46A
CR 210 mm
GW 24
Y94-62
Frontal
Section 541

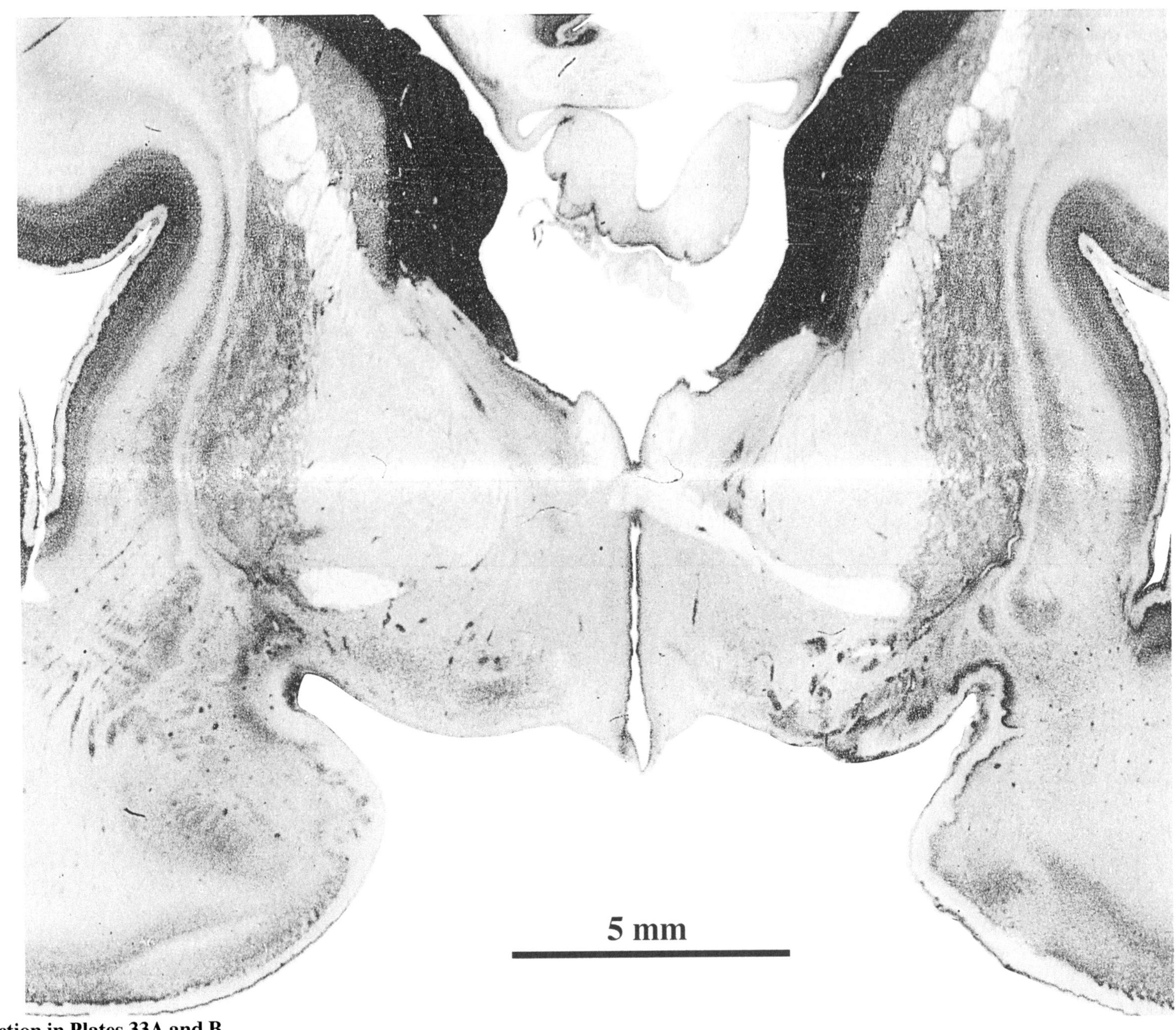

See the entire section in Plates 33A and B.

PLATE 46B

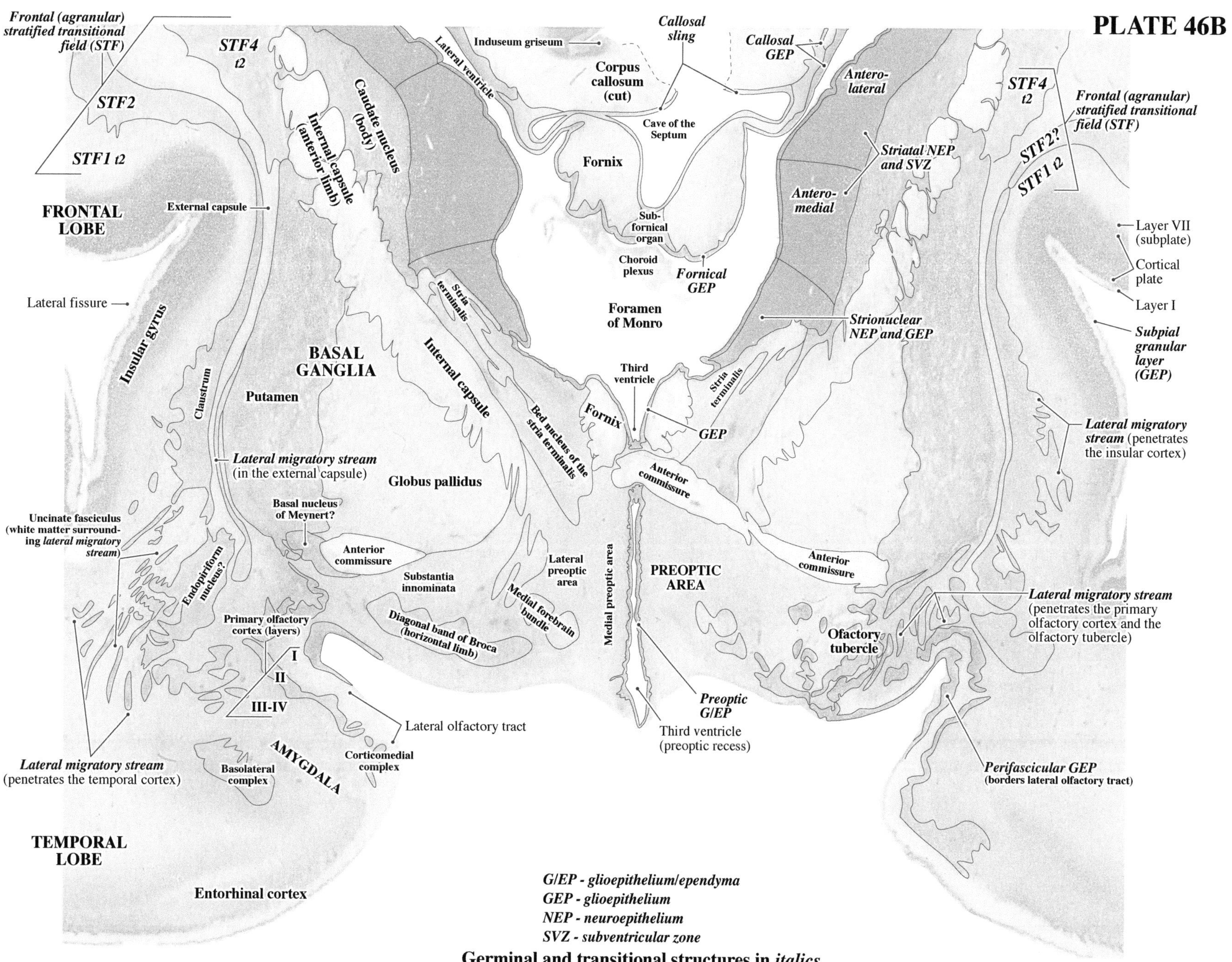

G/EP - glioepithelium/ependyma
GEP - glioepithelium
NEP - neuroepithelium
SVZ - subventricular zone

Germinal and transitional structures in *italics*

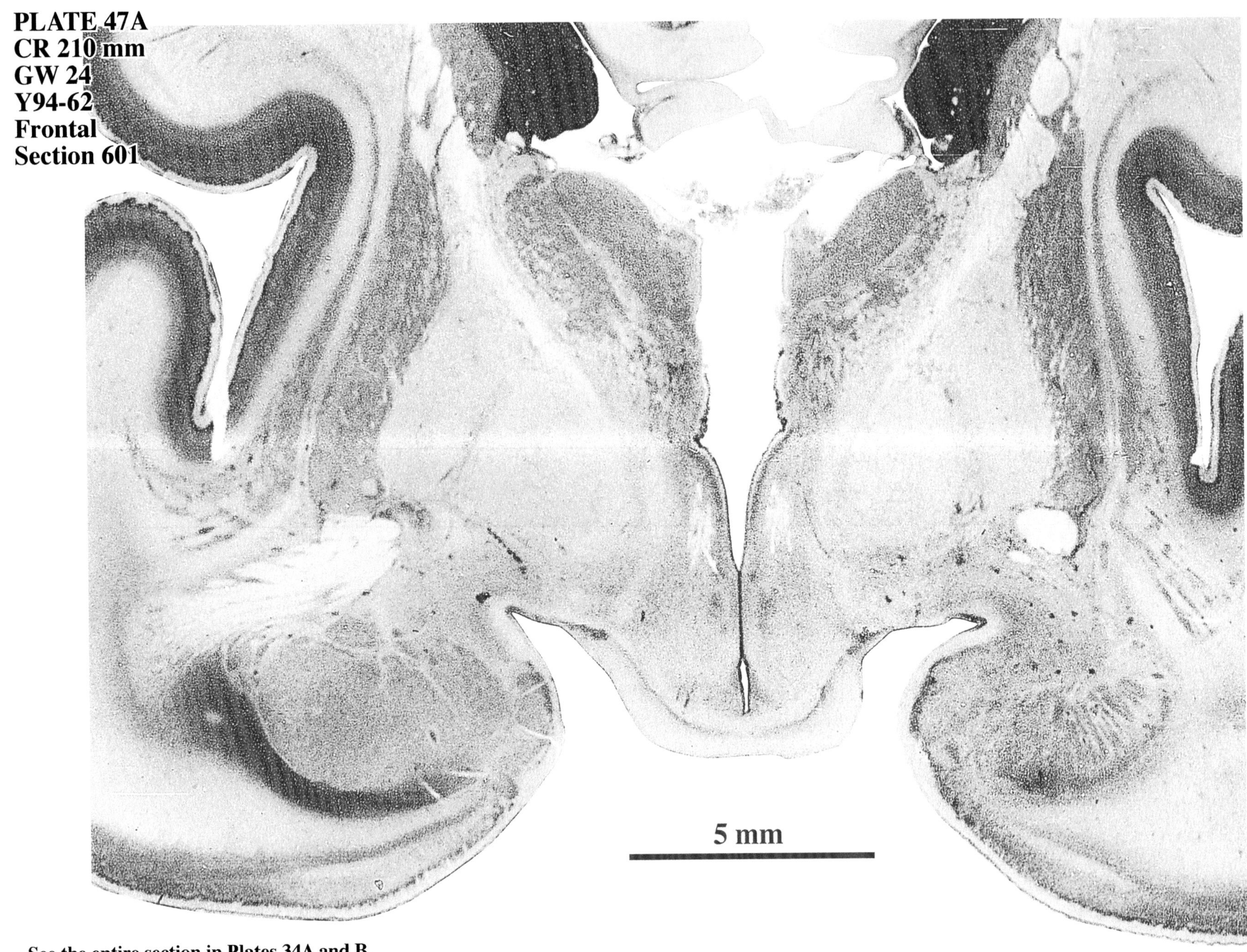

See the entire section in Plates 34A and B.

PLATE 47B

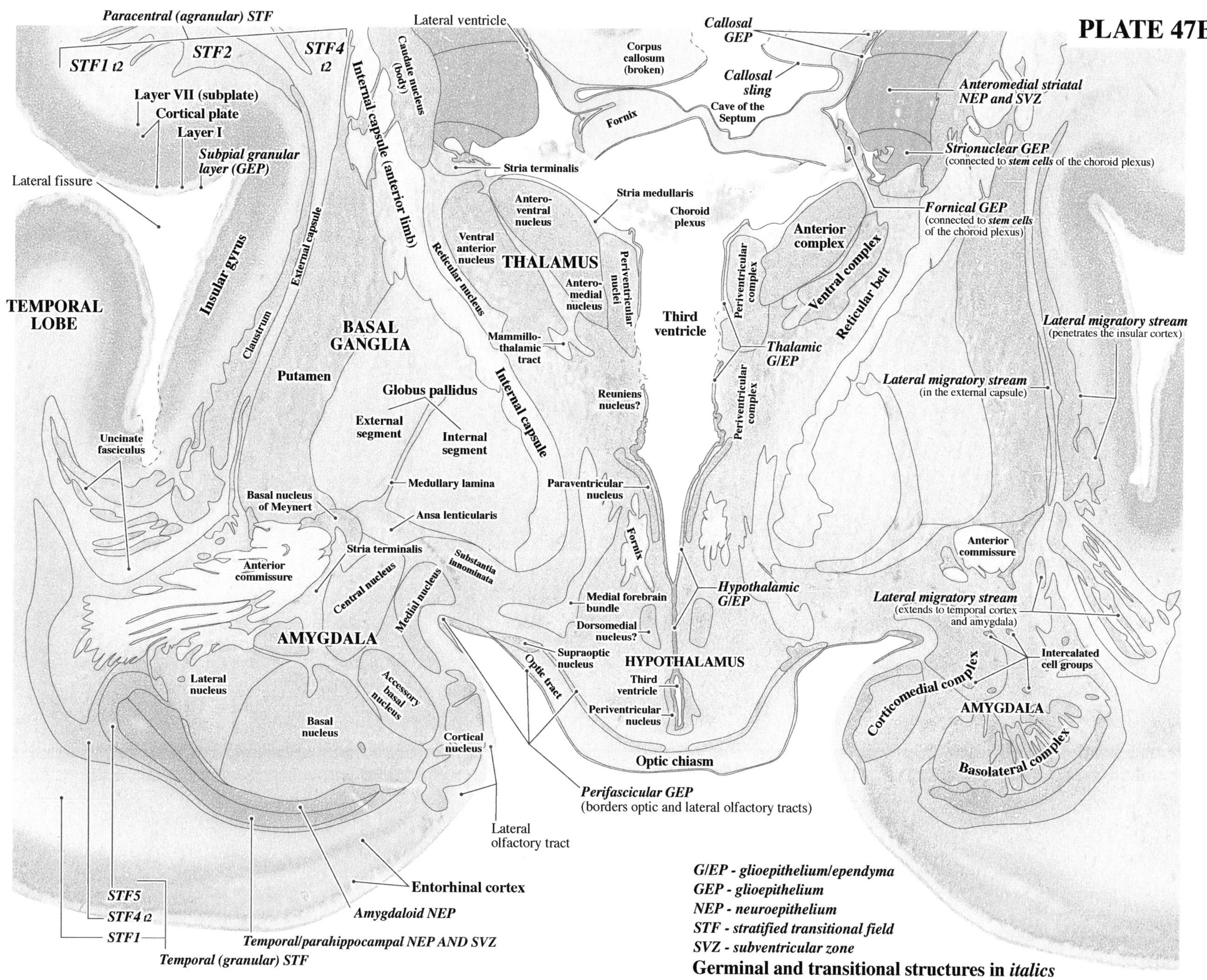

G/EP - *glioepithelium/ependyma*
GEP - *glioepithelium*
NEP - *neuroepithelium*
STF - *stratified transitional field*
SVZ - *subventricular zone*

Germinal and transitional structures in *italics*

PLATE 48A
CR 210 mm
GW 24
Y94-62
Frontal
Section 631

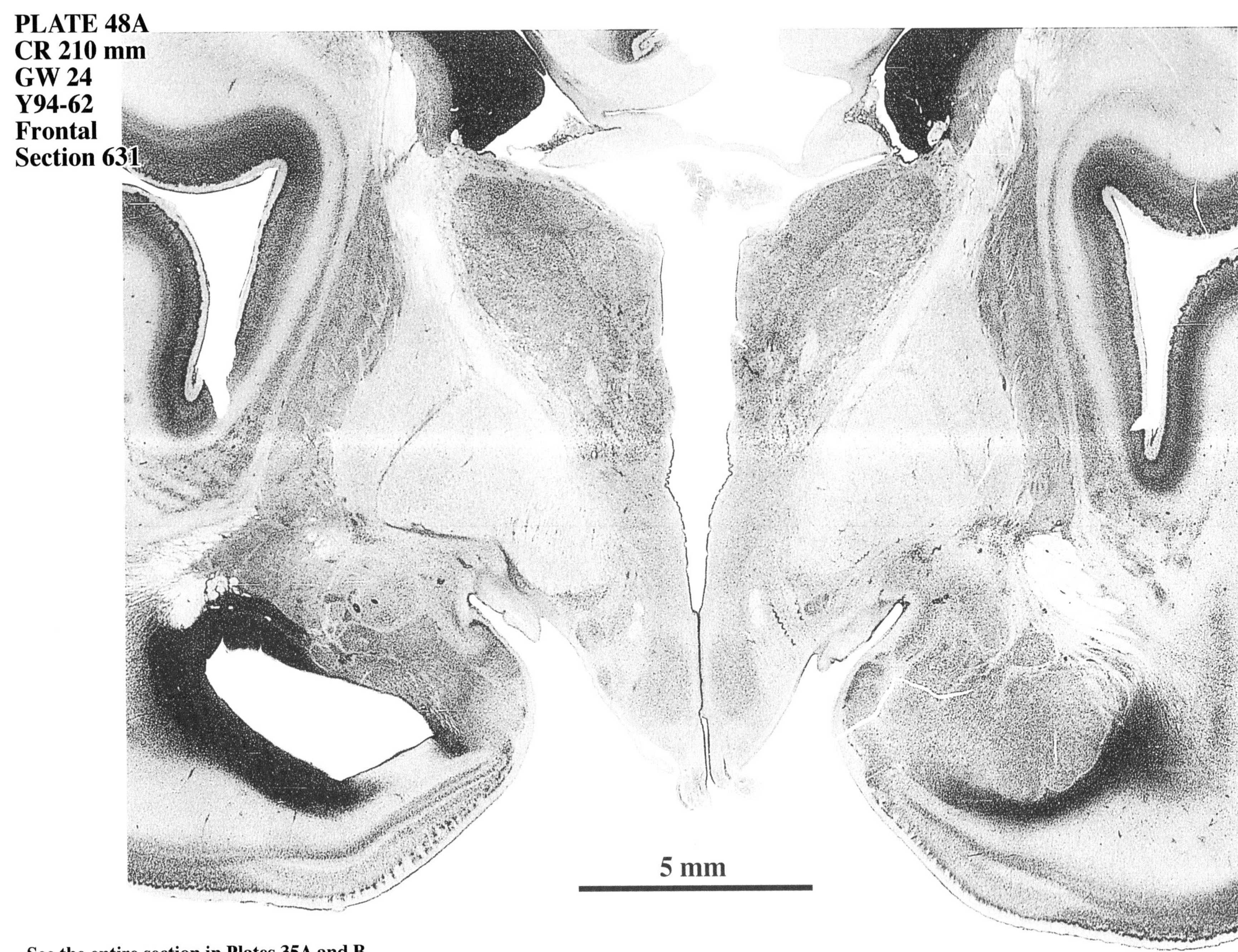

See the entire section in Plates 35A and B.

PLATE 48B

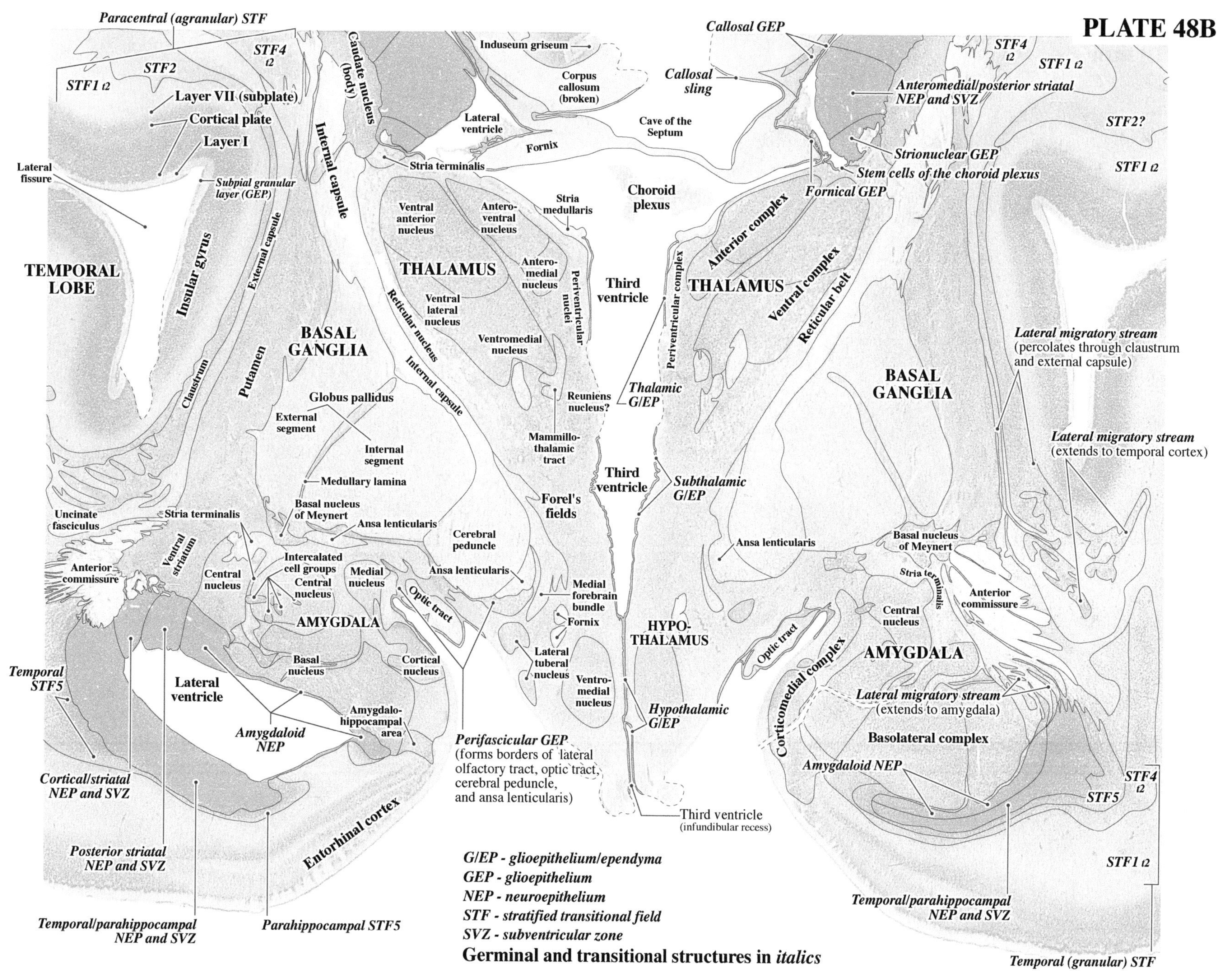

G/EP - glioepithelium/ependyma
GEP - glioepithelium
NEP - neuroepithelium
STF - stratified transitional field
SVZ - subventricular zone

Germinal and transitional structures in *italics*

PLATE 49A
CR 210 mm
GW 24
Y94-62
Frontal
Section 681

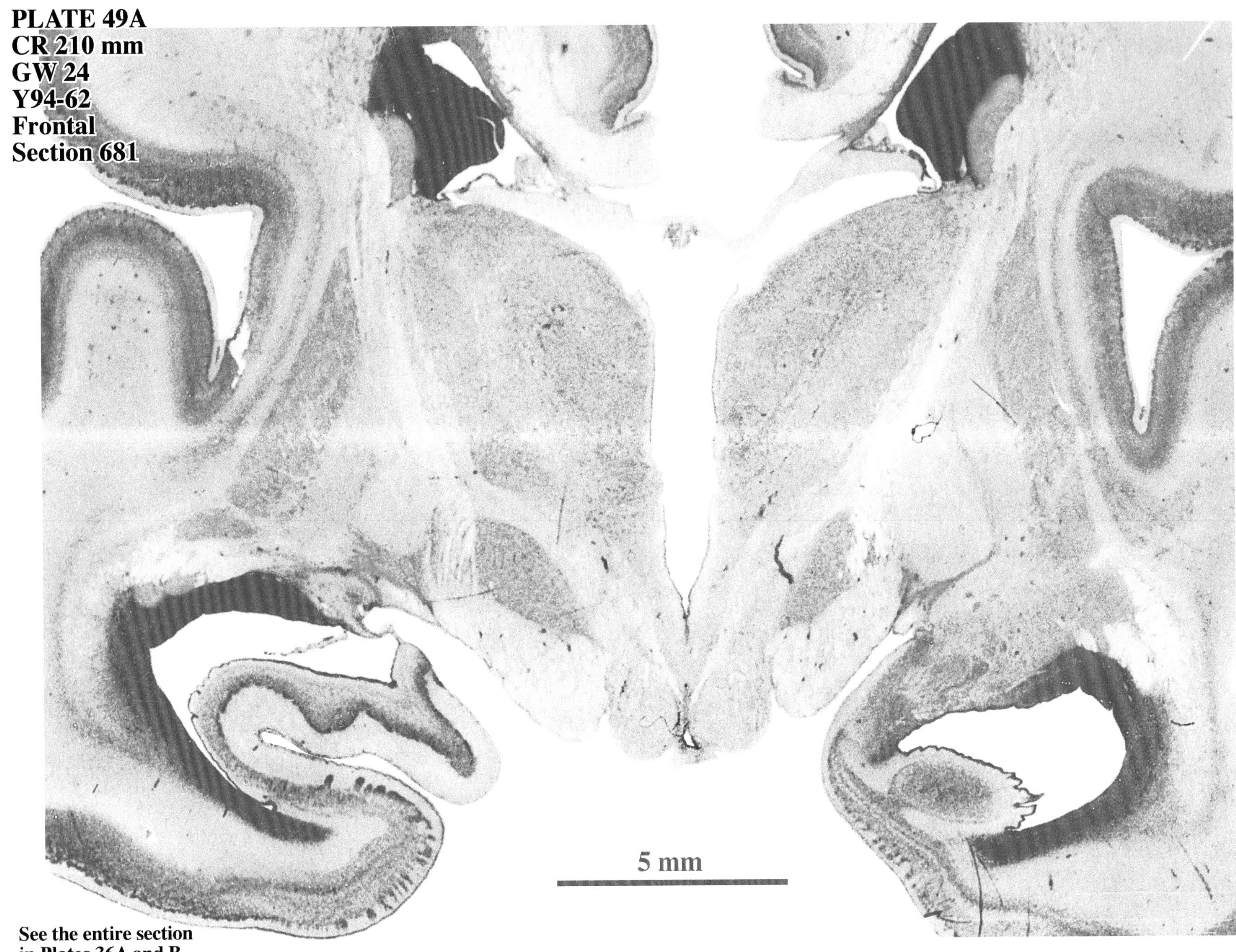

See the entire section in Plates 36A and B.

PLATE 49B

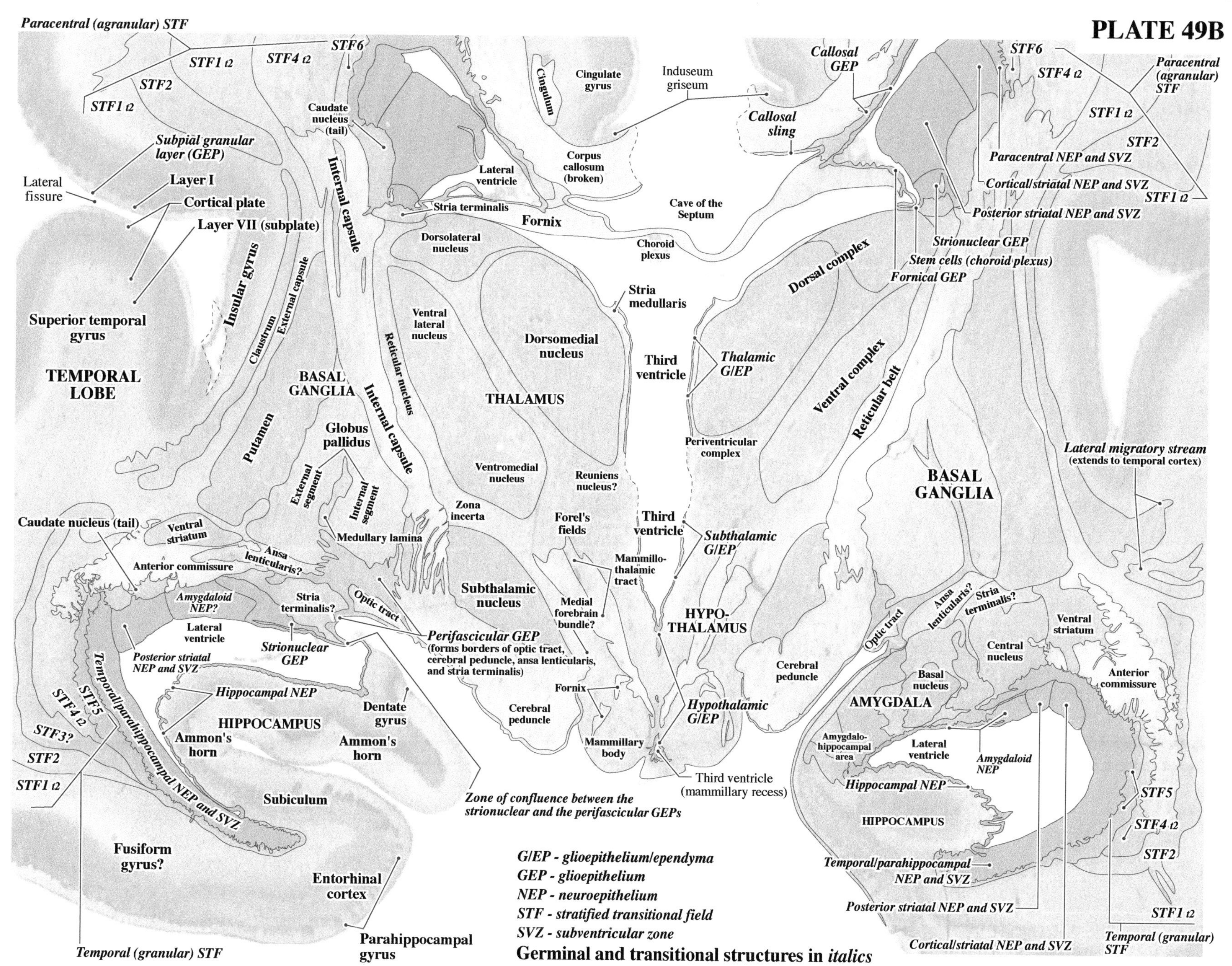

G/EP - glioepithelium/ependyma
GEP - glioepithelium
NEP - neuroepithelium
STF - stratified transitional field
SVZ - subventricular zone
Germinal and transitional structures in *italics*

PLATE 50A
CR 210 mm
GW 24
Y94-62
Frontal
Section 781

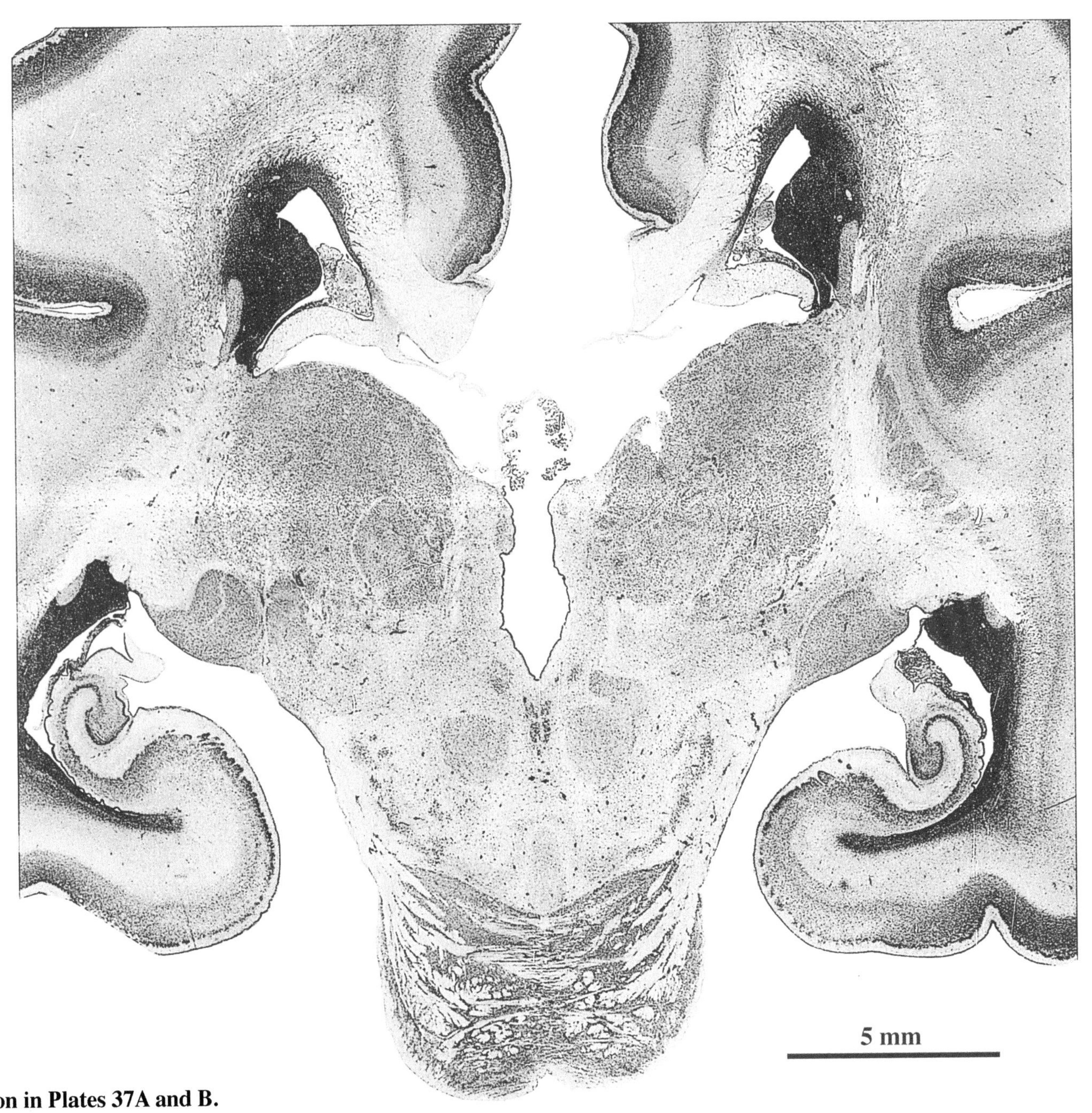

See the entire section in Plates 37A and B.

PLATE 50B

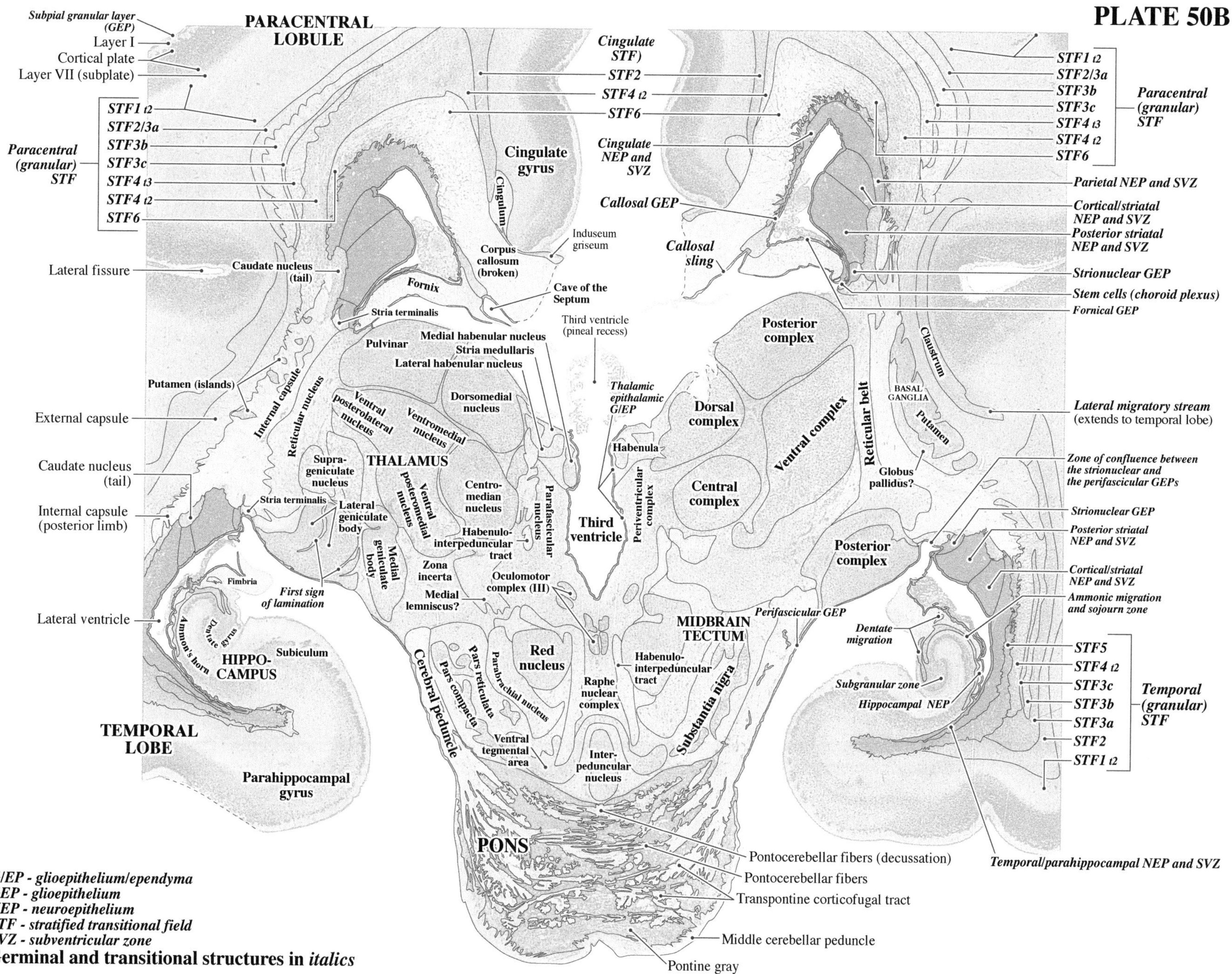

G/EP - glioepithelium/ependyma
GEP - glioepithelium
NEP - neuroepithelium
STF - stratified transitional field
SVZ - subventricular zone

Germinal and transitional structures in *italics*

PLATE 51A
CR 210 mm
GW 24
Y94-62
Frontal
Section 821

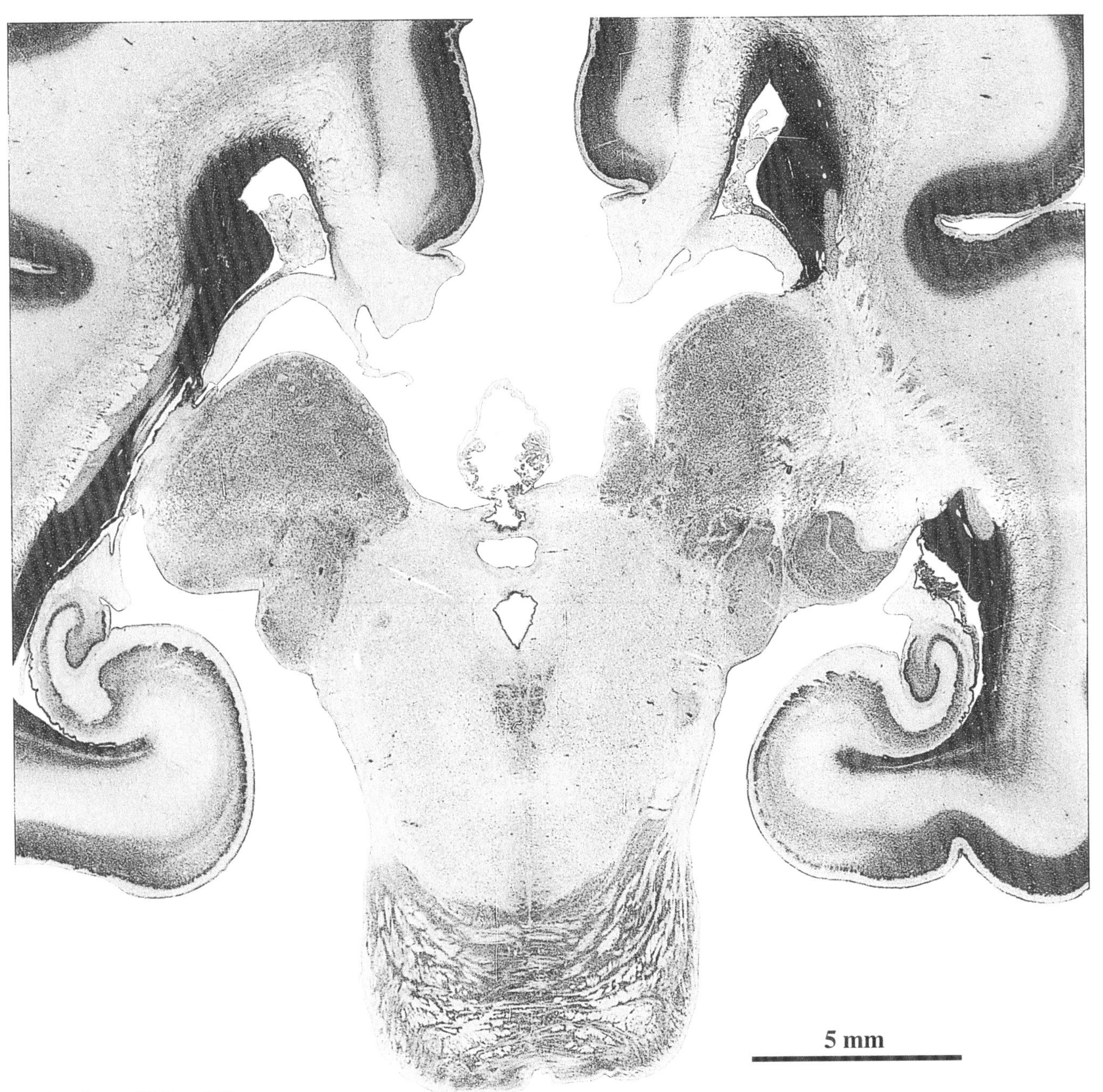

See the entire section in Plates 38A and B.

PLATE 51B

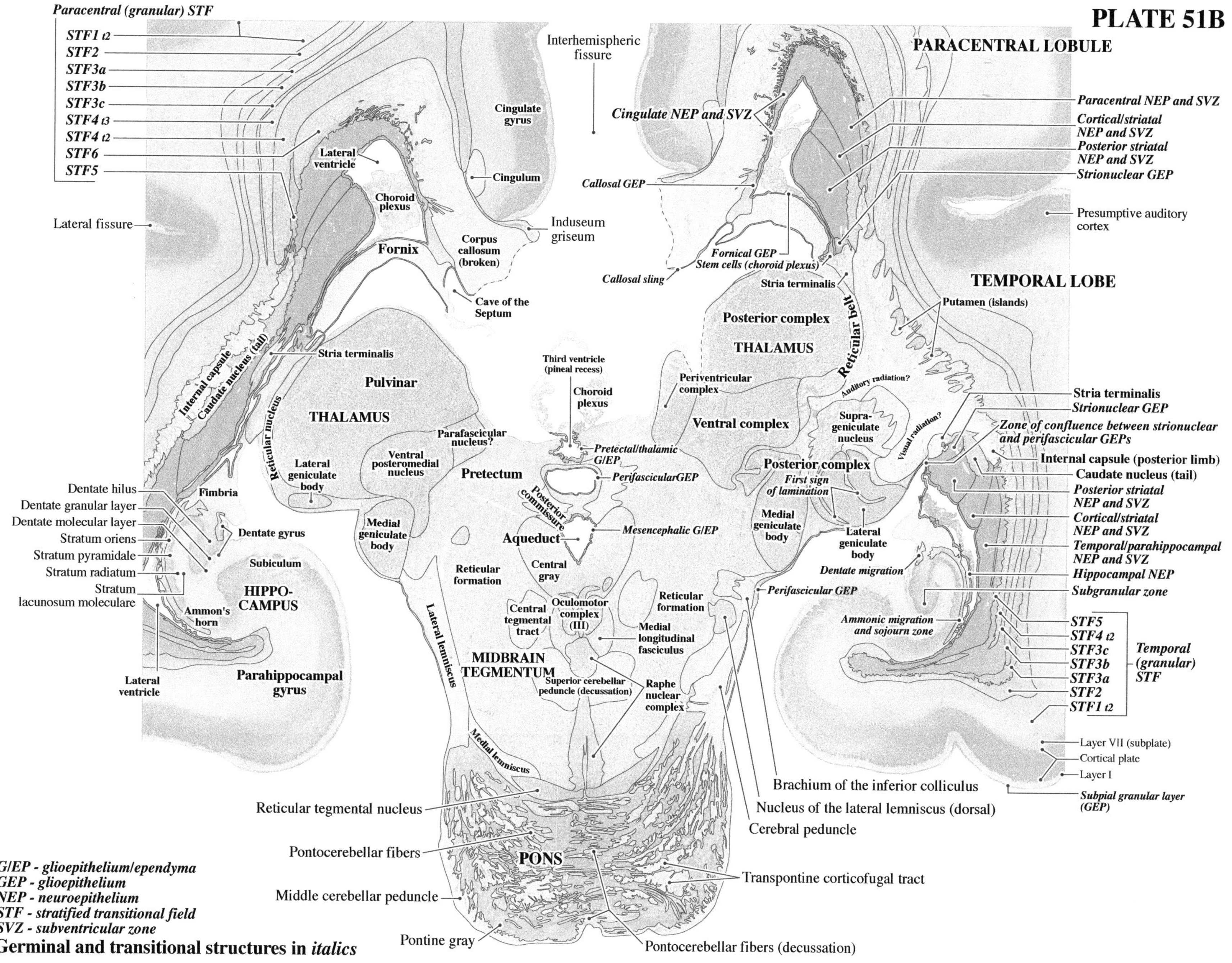
PARACENTRAL LOBULE
TEMPORAL LOBE
Paracentral (granular) STF
STF1 t2
STF2
STF3a
STF3b
STF3c
STF4 t3
STF4 t2
STF6
STF5
Lateral fissure
Lateral ventricle
Choroid plexus
Fornix
Cingulate gyrus
Cingulum
Interhemispheric fissure
Induseum griseum
Corpus callosum (broken)
Cave of the Septum
Cingulate NEP and SVZ
Callosal GEP
Callosal sling
Fornical GEP
Stem cells (choroid plexus)
Paracentral NEP and SVZ
Cortical/striatal NEP and SVZ
Posterior striatal NEP and SVZ
Strionuclear GEP
Presumptive auditory cortex
Stria terminalis
Reticular belt
Posterior complex
THALAMUS
Putamen (islands)
Internal capsule
Caudate nucleus (tail)
Reticular nucleus
Pulvinar
THALAMUS
Third ventricle (pineal recess)
Choroid plexus
Periventricular complex
Auditory radiation?
Ventral complex
Supra-geniculate nucleus
Visual radiation?
Stria terminalis
Strionuclear GEP
Zone of confluence between strionuclear and perifascicular GEPs
Internal capsule (posterior limb)
Caudate nucleus (tail)
Posterior striatal NEP and SVZ
Cortical/striatal NEP and SVZ
Temporal/parahippocampal NEP and SVZ
Hippocampal NEP
Subgranular zone
STF5
STF4 t2
STF3c
STF3b
STF3a
STF2
STF1 t2
Temporal (granular) STF
Layer VII (subplate)
Cortical plate
Layer I
Subpial granular layer (GEP)
Parafascicular nucleus?
Ventral posteromedial nucleus
Lateral geniculate body
Pretectum
Pretectal/thalamic G/EP
PerifascicularGEP
Posterior complex
First sign of lamination
Posterior commissure
Dentate hilus
Dentate granular layer
Dentate molecular layer
Stratum oriens
Stratum pyramidale
Stratum radiatum
Stratum lacunosum moleculare
Fimbria
Dentate gyrus
Subiculum
HIPPO-CAMPUS
Ammon's horn
Lateral ventricle
Parahippocampal gyrus
Medial geniculate body
Aqueduct
Mesencephalic G/EP
Medial geniculate body
Lateral geniculate body
Dentate migration
Perifascicular GEP
Reticular formation
Central gray
Reticular formation
Central tegmental tract
Oculomotor complex (III)
Medial longitudinal fasciculus
Ammonic migration and sojourn zone
Lateral lemniscus
MIDBRAIN TEGMENTUM
Superior cerebellar peduncle (decussation)
Raphe nuclear complex
Medial lemniscus
Brachium of the inferior colliculus
Nucleus of the lateral lemniscus (dorsal)
Cerebral peduncle
Reticular tegmental nucleus
Pontocerebellar fibers
PONS
Transpontine corticofugal tract
Middle cerebellar peduncle
Pontine gray
Pontocerebellar fibers (decussation)
G/EP - glioepithelium/ependyma
GEP - glioepithelium
NEP - neuroepithelium
STF - stratified transitional field
SVZ - subventricular zone
Germinal and transitional structures in italics

PLATE 52A
CR 210 mm
GW 24
Y94-62
Frontal
Section 861

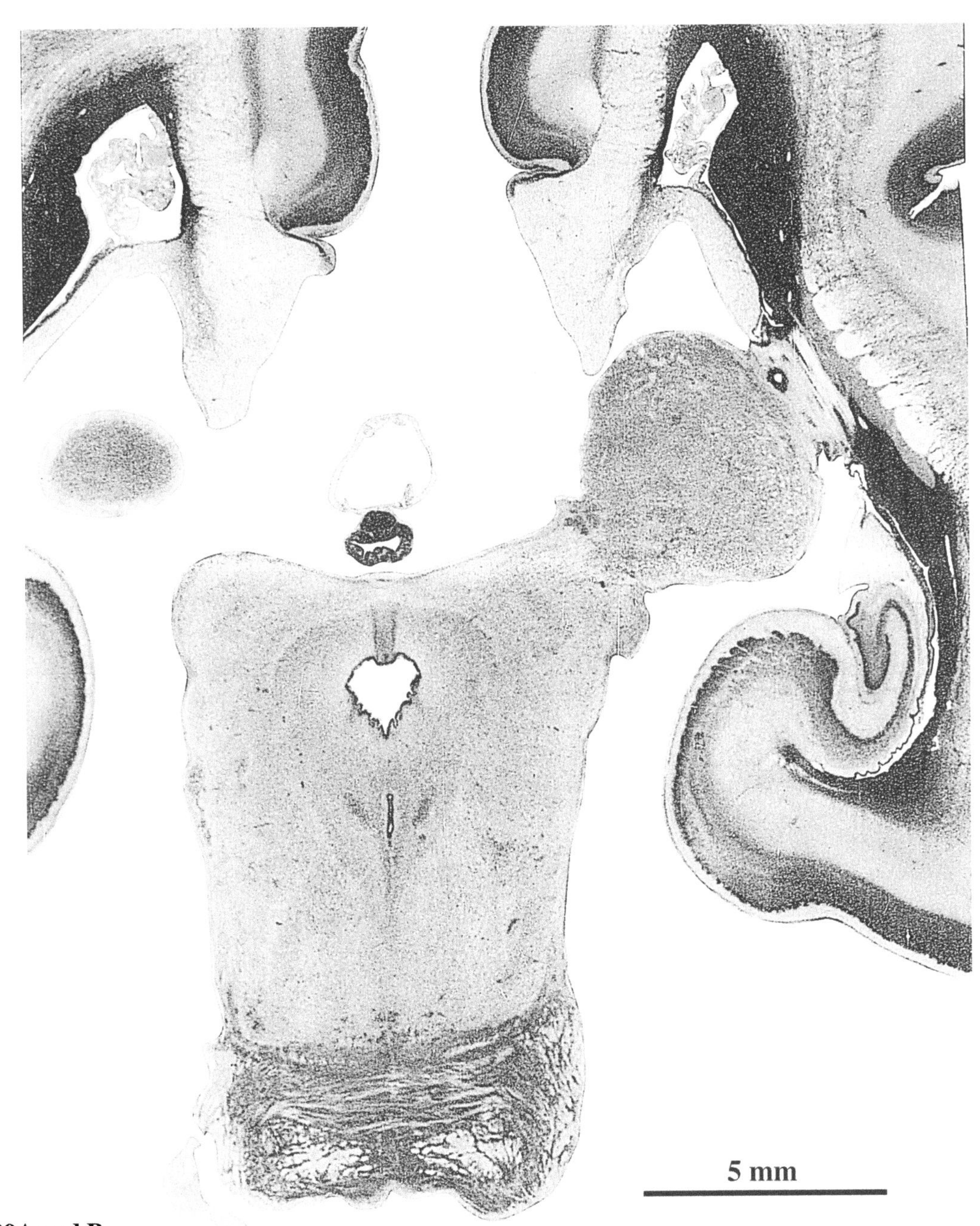

See the entire section in Plates 39A and B.

PLATE 52B

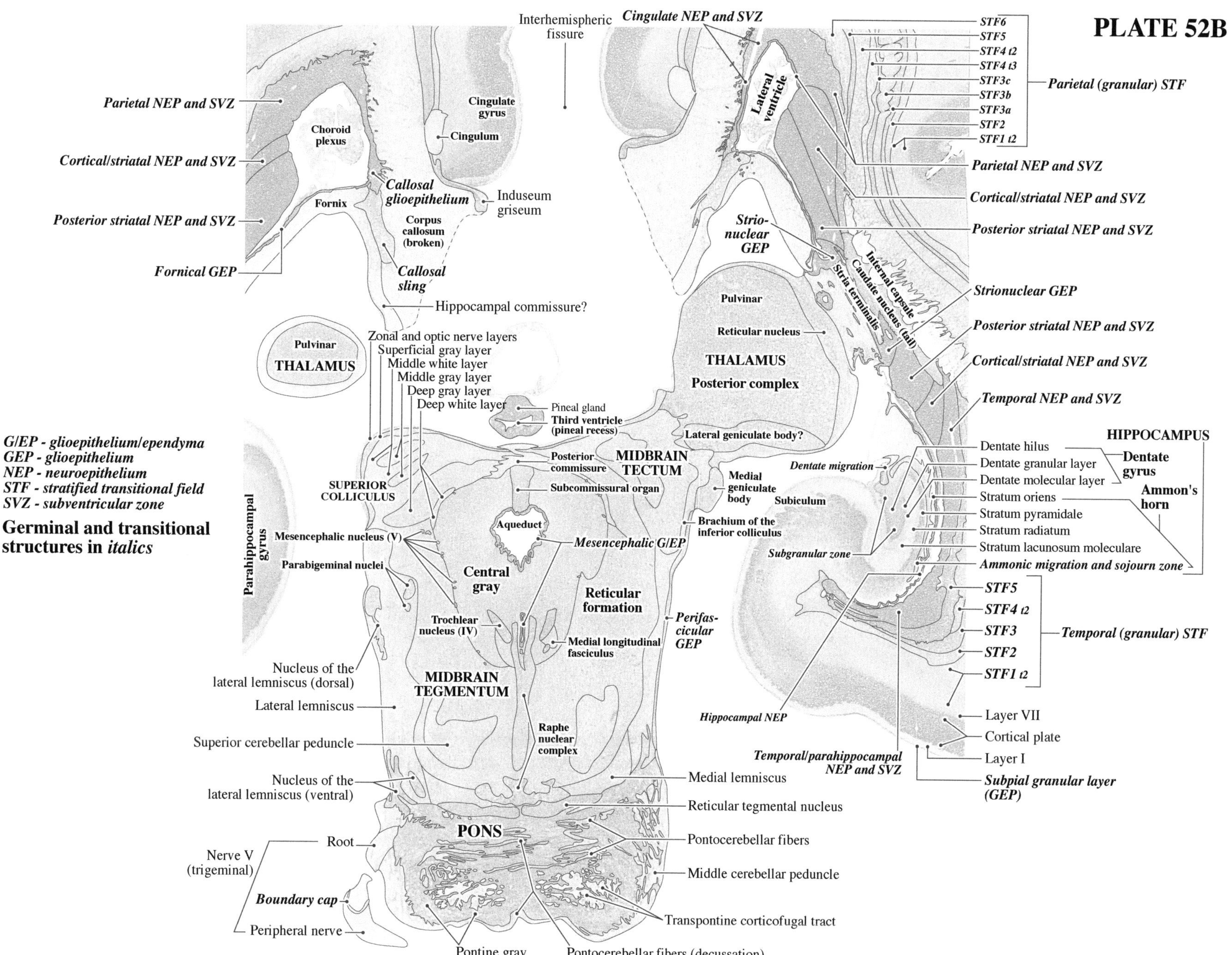

G/EP - glioepithelium/ependyma
GEP - glioepithelium
NEP - neuroepithelium
STF - stratified transitional field
SVZ - subventricular zone

Germinal and transitional structures in *italics*

PLATE 53A
CR 210 mm
GW 24
Y94-62
Frontal
Section 901

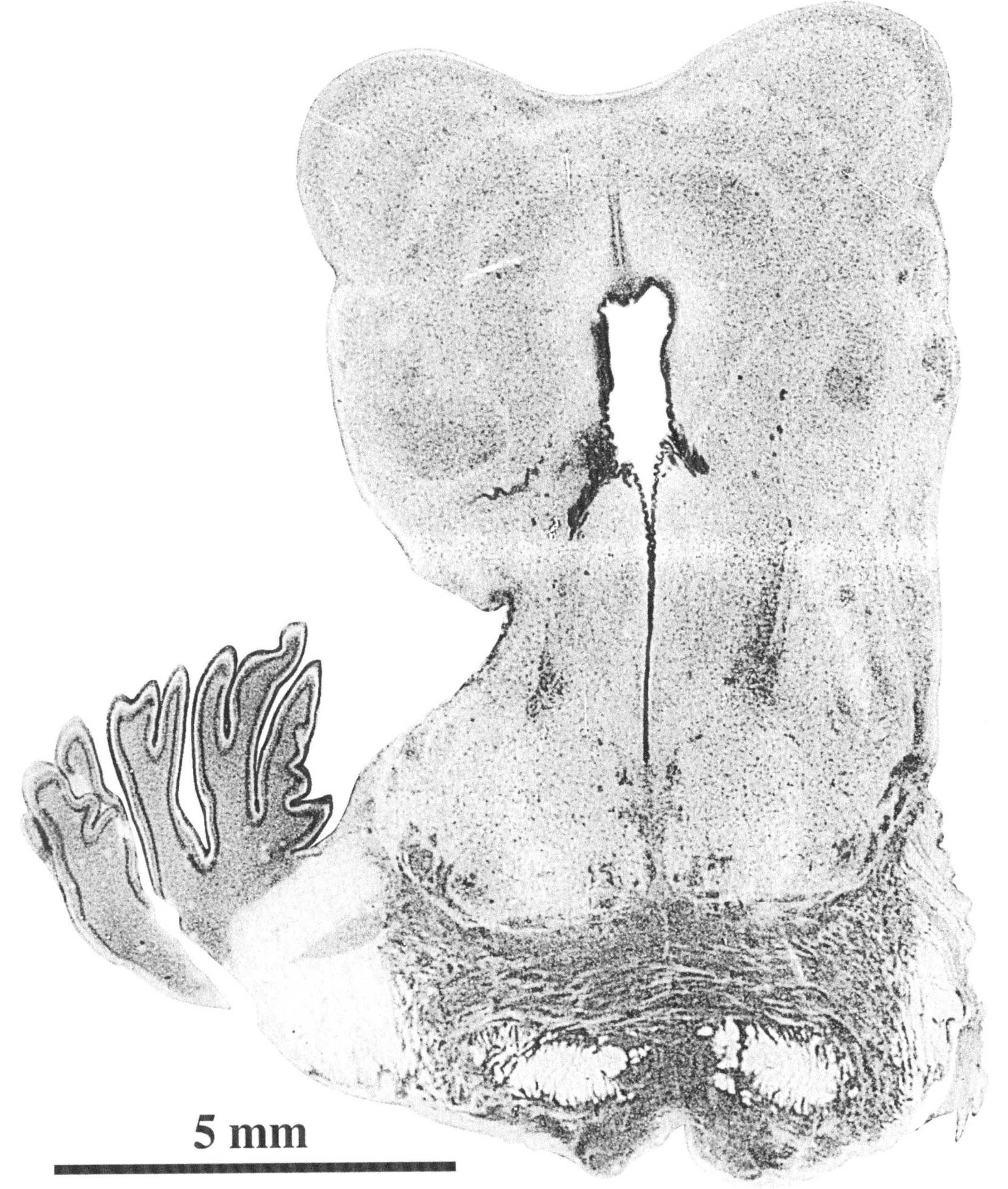

See the entire section in Plates 40A and B.

PLATE 53B

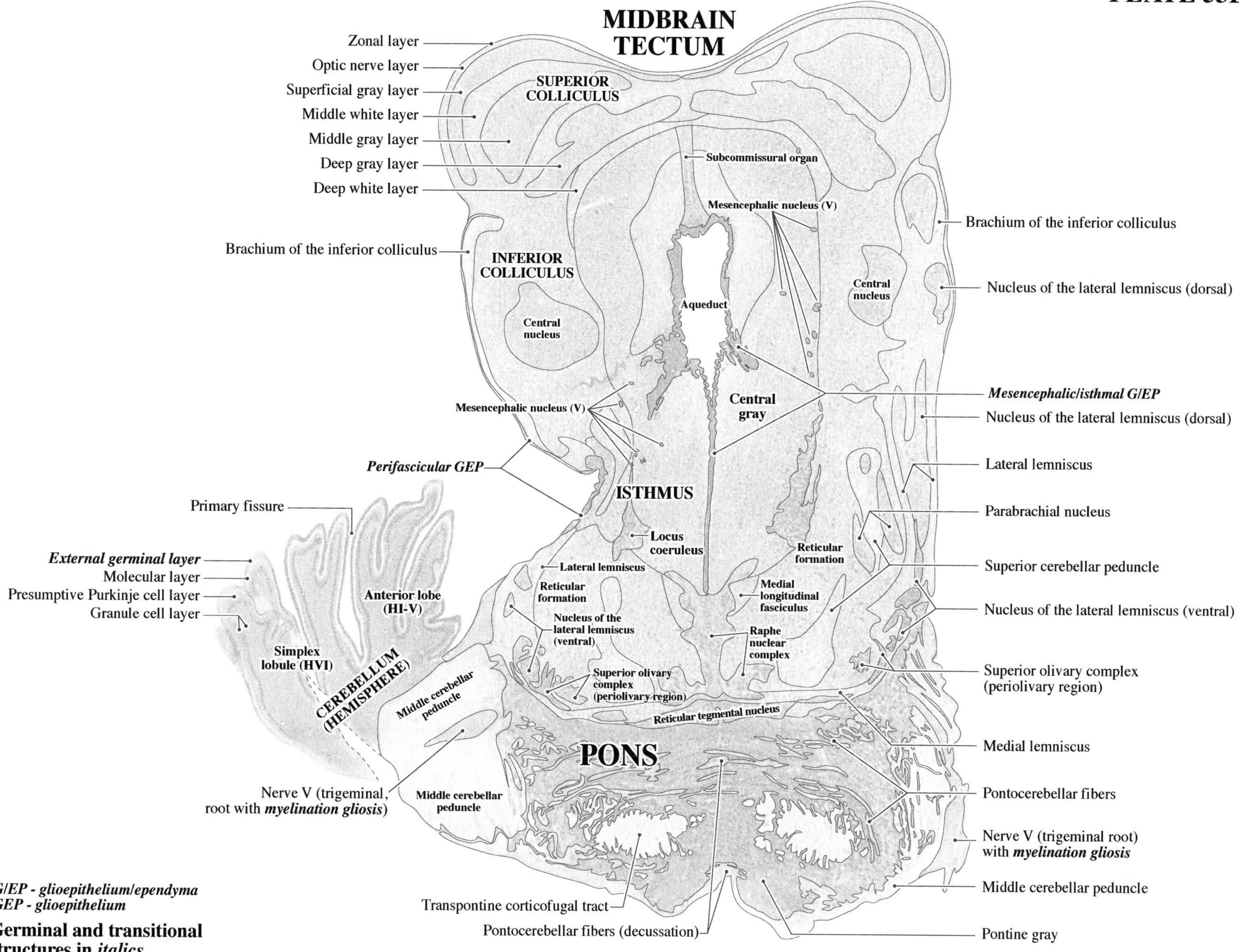
MIDBRAIN
TECTUM
Zonal layer
Optic nerve layer
Superficial gray layer
Middle white layer
Middle gray layer
Deep gray layer
Deep white layer
SUPERIOR
COLLICULUS
Subcommissural organ
Mesencephalic nucleus (V)
Brachium of the inferior colliculus
Brachium of the inferior colliculus
INFERIOR
COLLICULUS
Central
nucleus
Aqueduct
Central
nucleus
Nucleus of the lateral lemniscus (dorsal)
Mesencephalic/isthmal G/EP
Central
gray
Mesencephalic nucleus (V)
Nucleus of the lateral lemniscus (dorsal)
Perifascicular GEP
Lateral lemniscus
ISTHMUS
Primary fissure
Parabrachial nucleus
Locus
coeruleus
Reticular
formation
External germinal layer
Molecular layer
Presumptive Purkinje cell layer
Granule cell layer
Lateral lemniscus
Superior cerebellar peduncle
Anterior lobe
(HI-V)
Reticular
formation
Medial
longitudinal
fasciculus
Nucleus of the lateral lemniscus (ventral)
Nucleus of the
lateral lemniscus
(ventral)
Raphe
nuclear
complex
Simplex
lobule (HVI)
CEREBELLUM
(HEMISPHERE)
Superior olivary
complex
(periolivary region)
Superior olivary complex
(periolivary region)
Middle cerebellar
peduncle
Reticular tegmental nucleus
Medial lemniscus
PONS
Pontocerebellar fibers
Nerve V (trigeminal,
root with myelination gliosis)
Middle cerebellar
peduncle
Nerve V (trigeminal root)
with myelination gliosis
Middle cerebellar peduncle
Transpontine corticofugal tract
Pontocerebellar fibers (decussation)
Pontine gray
G/EP - glioepithelium/ependyma
GEP - glioepithelium
Germinal and transitional
structures in italics

PLATE 54A
CR 210 mm
GW 24
Y94-62
Frontal
Section 941

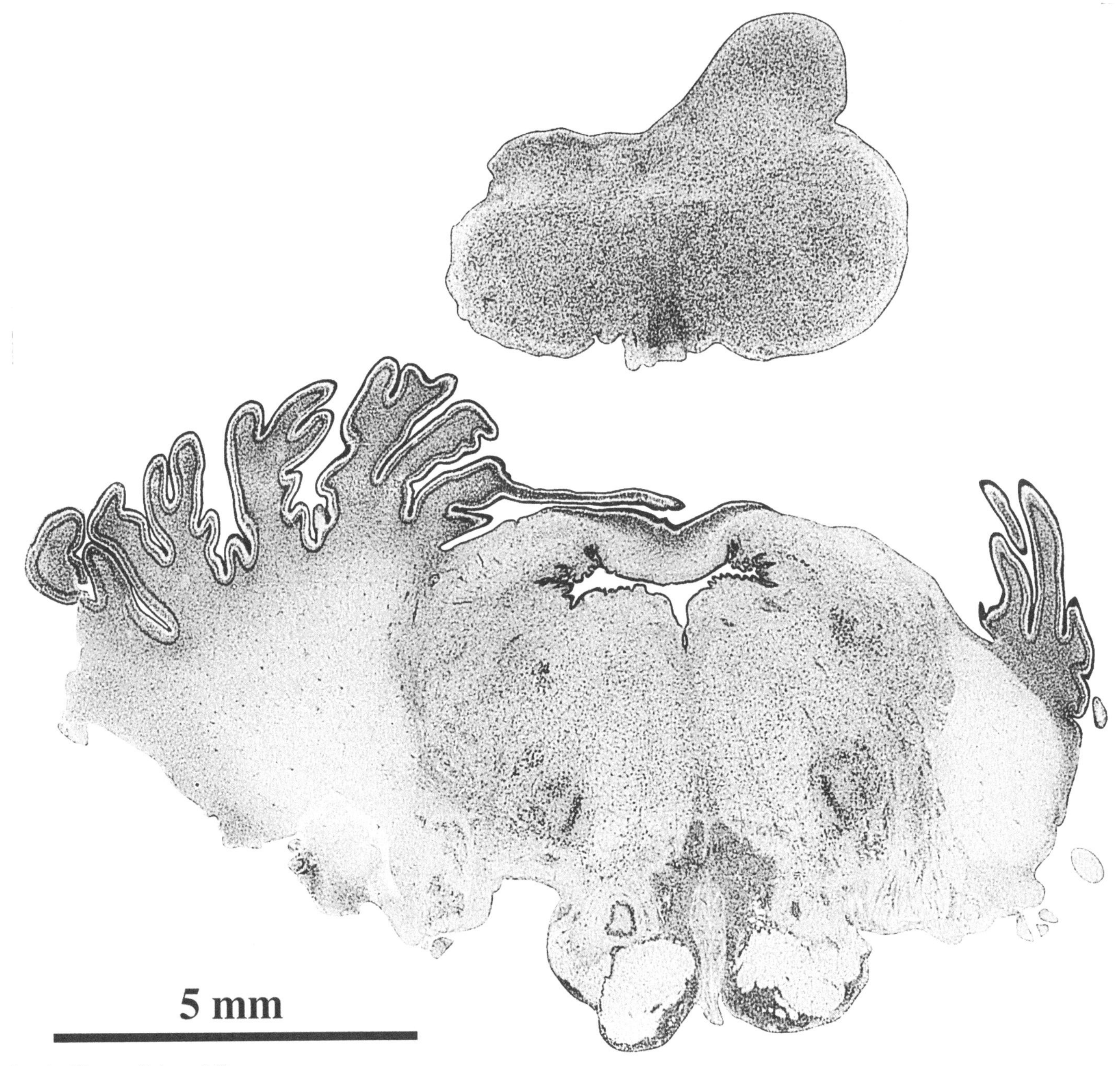

See the entire section in Plates 41A and B.

PLATE 54B

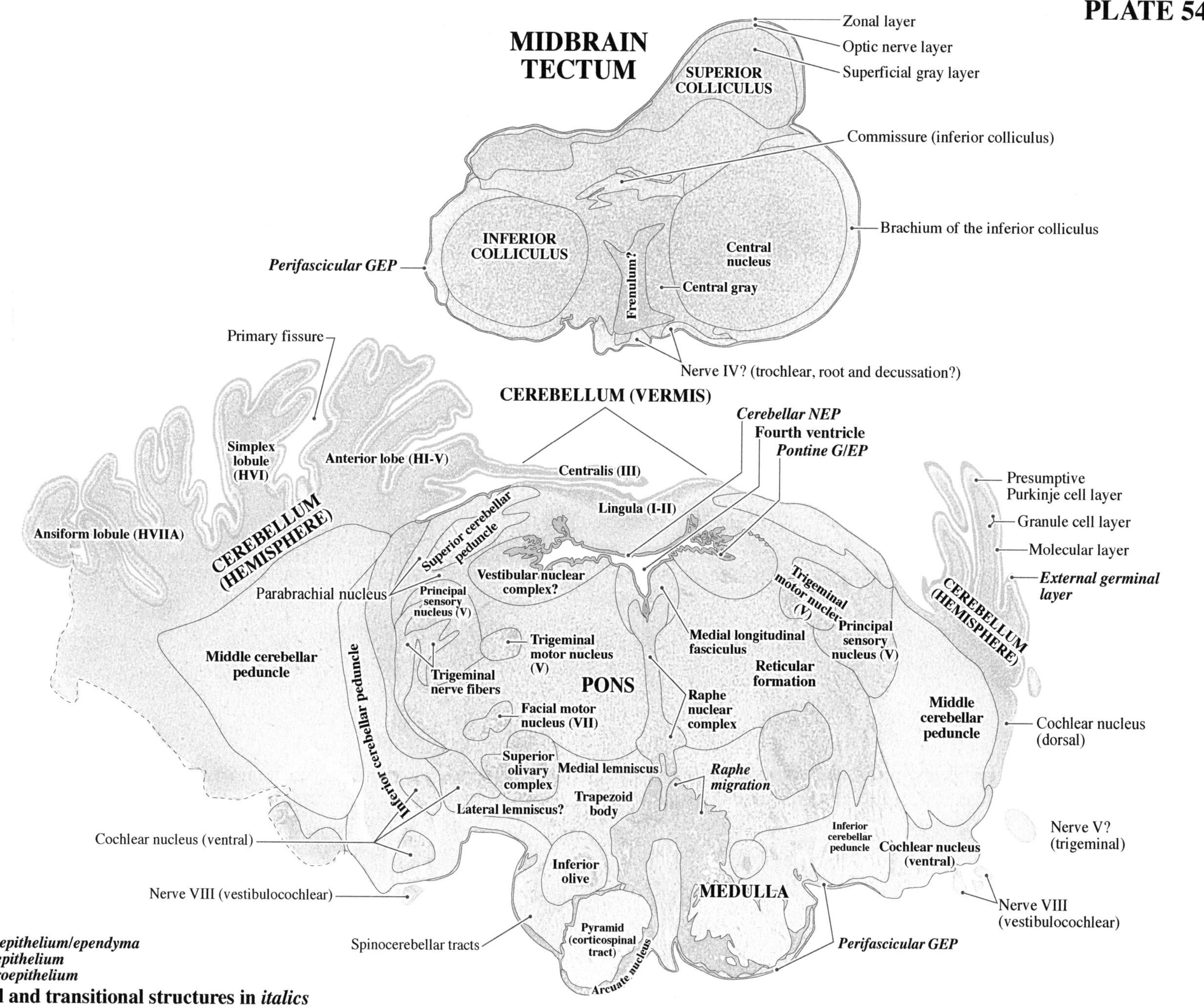
MIDBRAIN TECTUM
SUPERIOR COLLICULUS
Zonal layer
Optic nerve layer
Superficial gray layer
Commissure (inferior colliculus)
Brachium of the inferior colliculus
INFERIOR COLLICULUS
Perifascicular GEP
Frenulum?
Central nucleus
Central gray
Nerve IV? (trochlear, root and decussation?)
Primary fissure
CEREBELLUM (VERMIS)
Cerebellar NEP
Fourth ventricle
Pontine G/EP
Simplex lobule (HVI)
Anterior lobe (HI-V)
Centralis (III)
Lingula (I-II)
Ansiform lobule (HVIIA)
CEREBELLUM (HEMISPHERE)
Superior cerebellar peduncle
Vestibular nuclear complex?
Parabrachial nucleus
Principal sensory nucleus (V)
Trigeminal motor nuclei (V)
Principal sensory nucleus (V)
Presumptive Purkinje cell layer
Granule cell layer
Molecular layer
External germinal layer
CEREBELLUM (HEMISPHERE)
Middle cerebellar peduncle
Inferior cerebellar peduncle
Trigeminal motor nucleus (V)
Trigeminal nerve fibers
Medial longitudinal fasciculus
Reticular formation
PONS
Raphe nuclear complex
Facial motor nucleus (VII)
Middle cerebellar peduncle
Cochlear nucleus (dorsal)
Superior olivary complex
Medial lemniscus
Raphe migration
Trapezoid body
Lateral lemniscus?
Nerve V? (trigeminal)
Cochlear nucleus (ventral)
Inferior cerebellar peduncle
Cochlear nucleus (ventral)
Inferior olive
MEDULLA
Nerve VIII (vestibulocochlear)
Nerve VIII (vestibulocochlear)
Pyramid (corticospinal tract)
Spinocerebellar tracts
Arcuate nucleus
Perifascicular GEP
G/EP - glioepithelium/ependyma
GEP - glioepithelium
NEP - neuroepithelium
Germinal and transitional structures in italics

PLATE 55A
CR 210 mm
GW 24
Y94-62
Frontal
Section 981

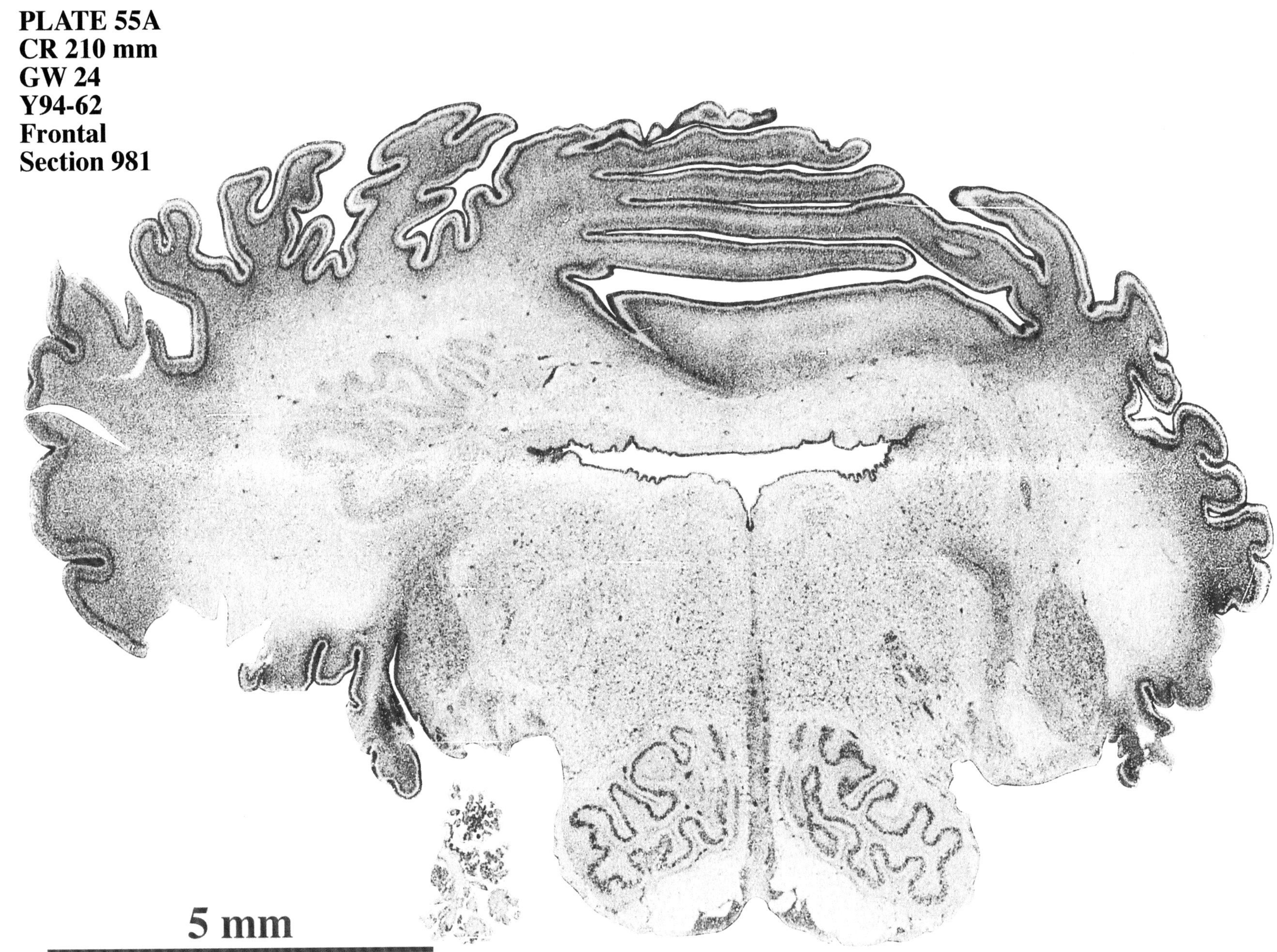

See the entire section in Plates 42A and B.

PLATE 55B

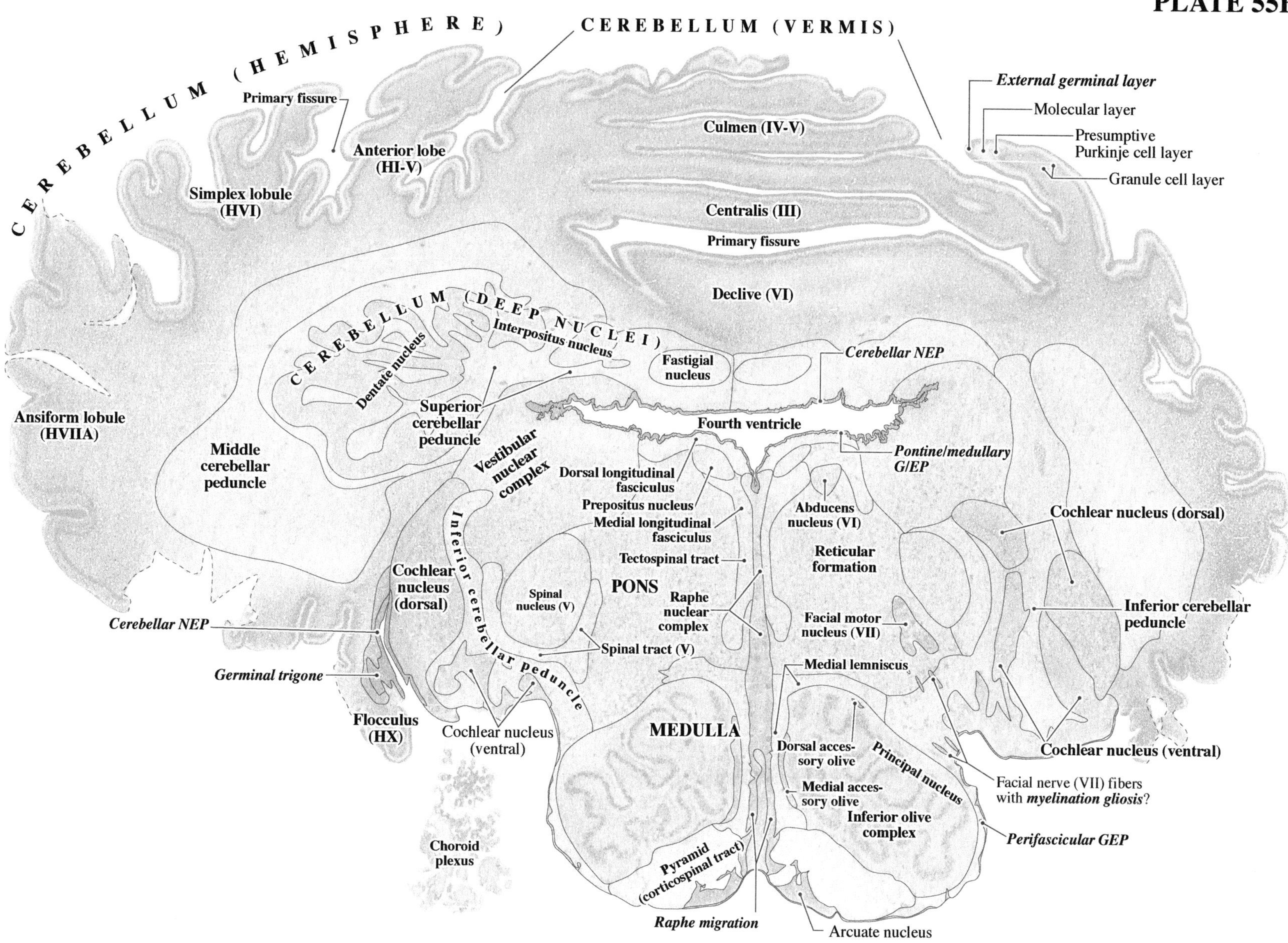

G/EP - glioepithelium/ependyma
GEP - glioepithelium
NEP - neuroepithelium

Germinal and transitional structures in *italics*

PLATE 56A
CR 210 mm
GW 24, Y94-62
Frontal, Section 1041

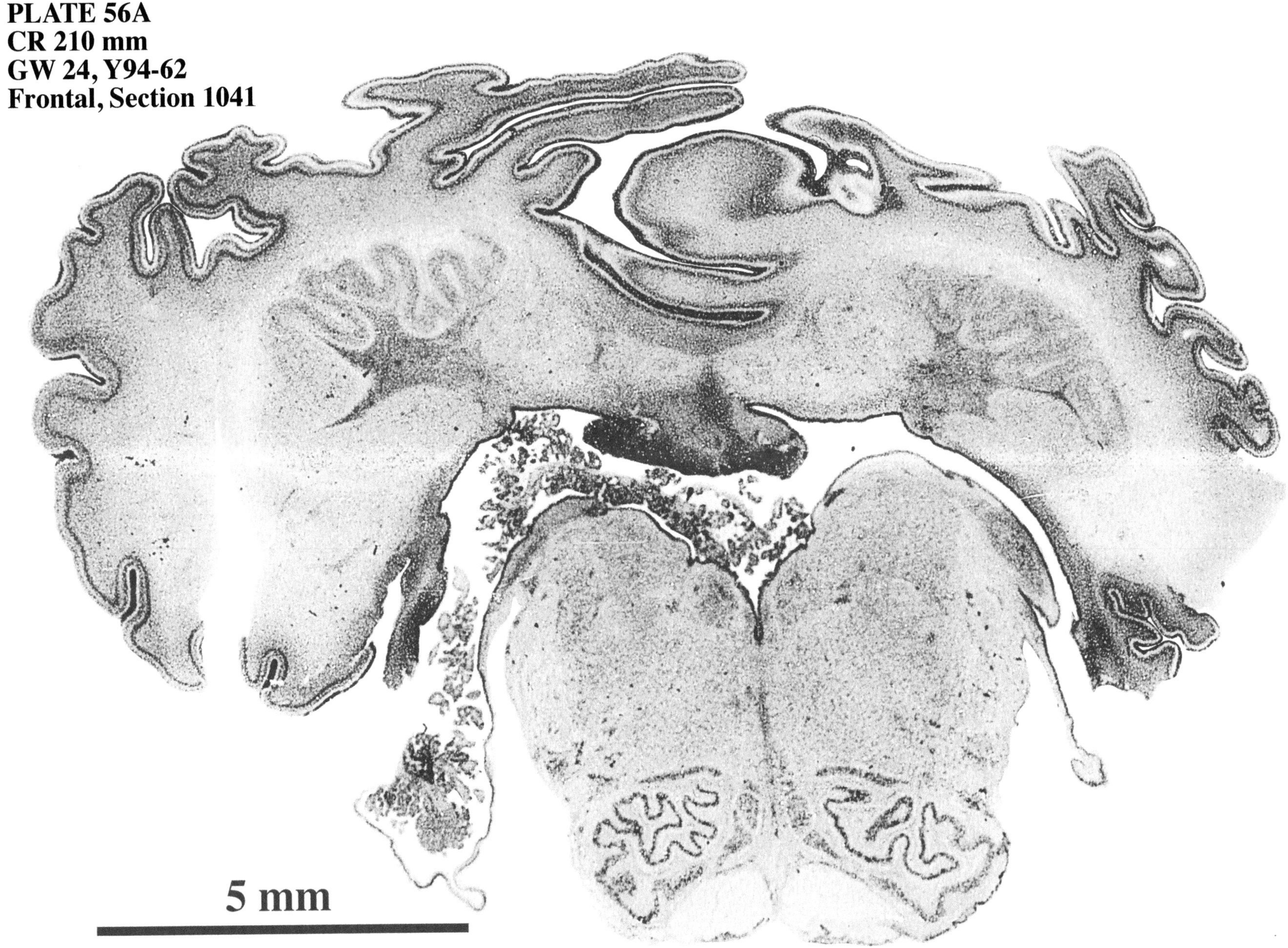

See the entire section in Plates 43A and B.

PLATE 56B

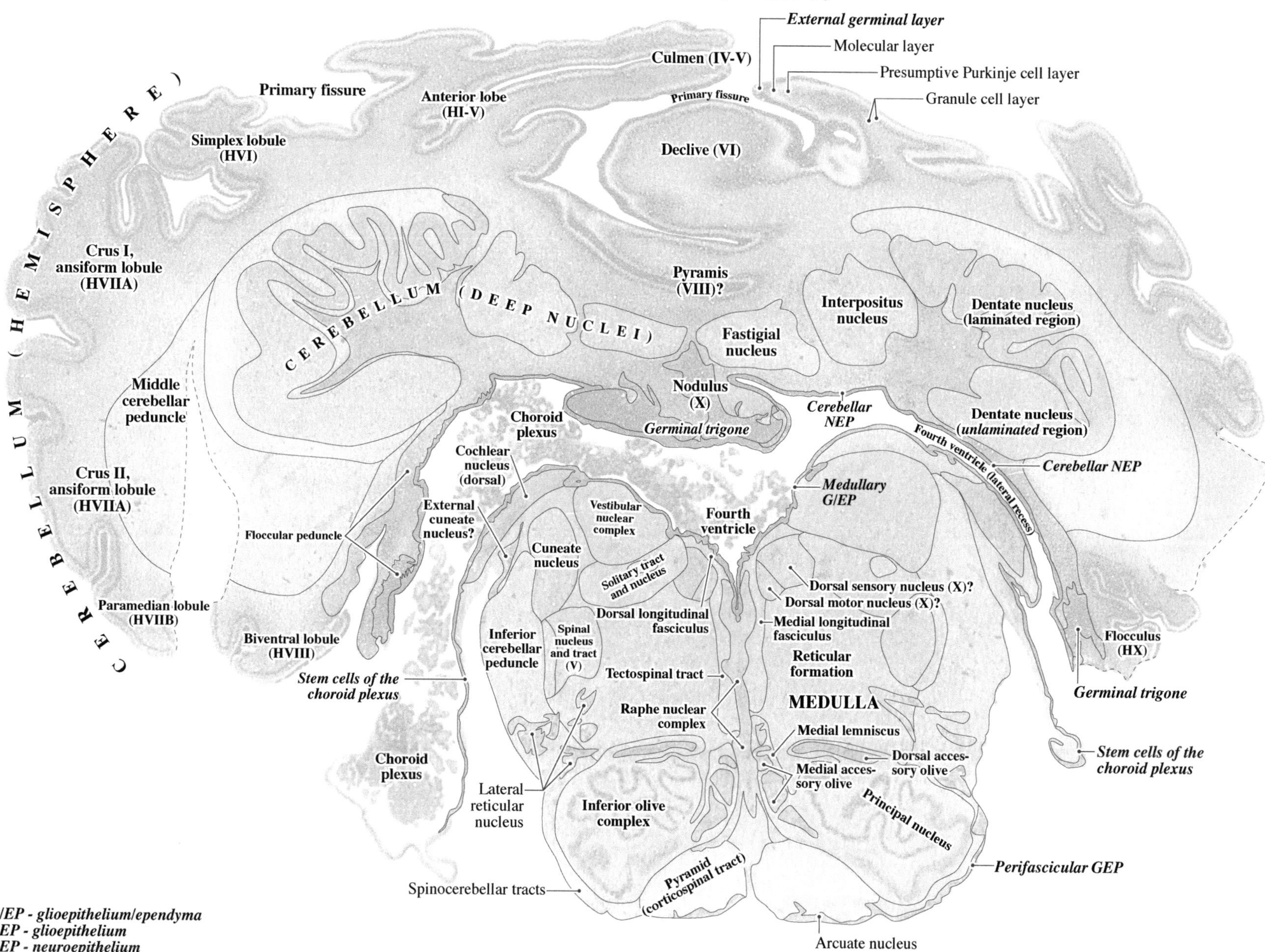

G/EP - glioepithelium/ependyma
GEP - glioepithelium
NEP - neuroepithelium

Germinal and transitional structures in *italics*

PLATE 57A
CR 210 mm
GW 24, Y94-62
Frontal
Section 1141

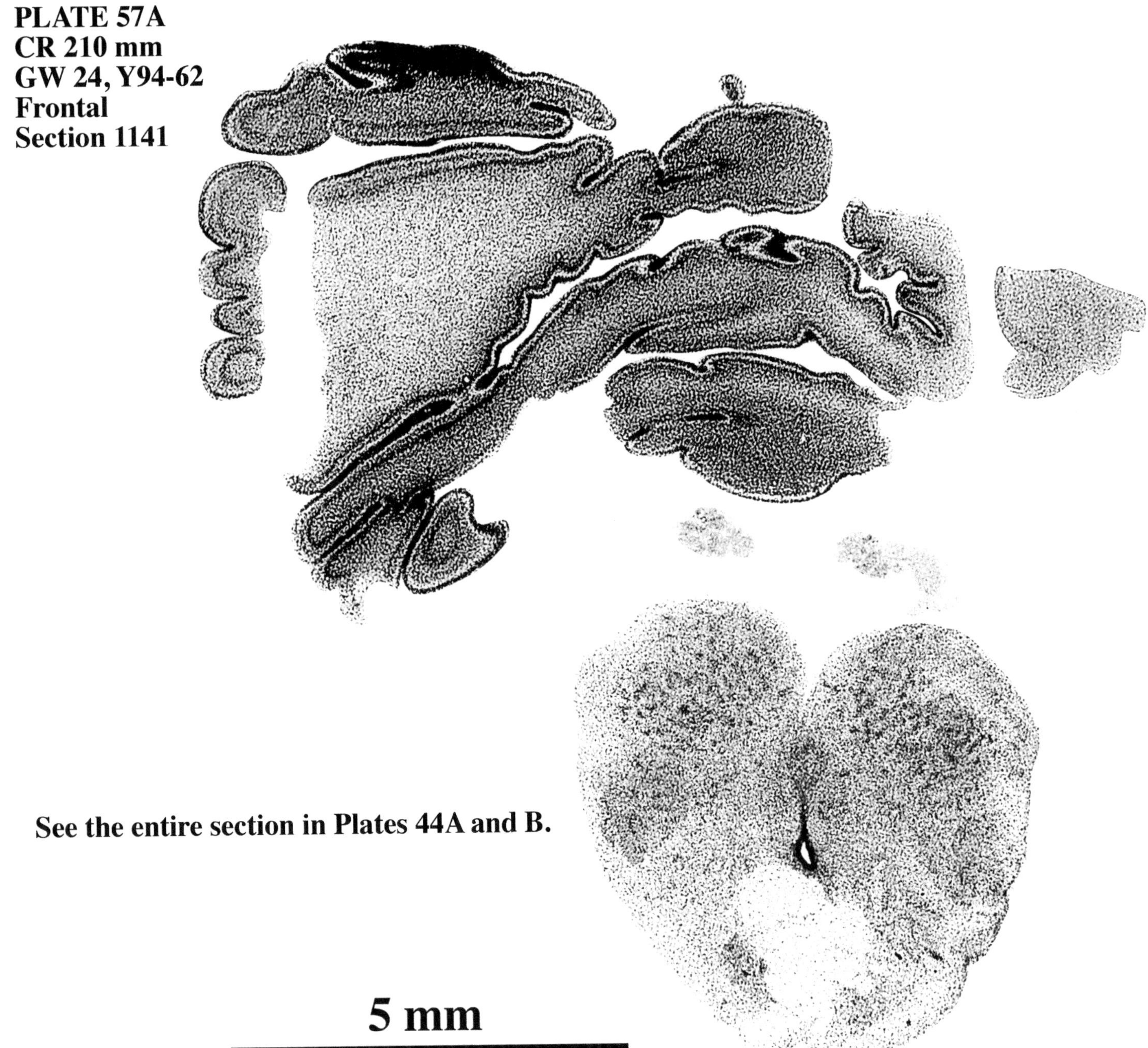

See the entire section in Plates 44A and B.

PLATE 57B

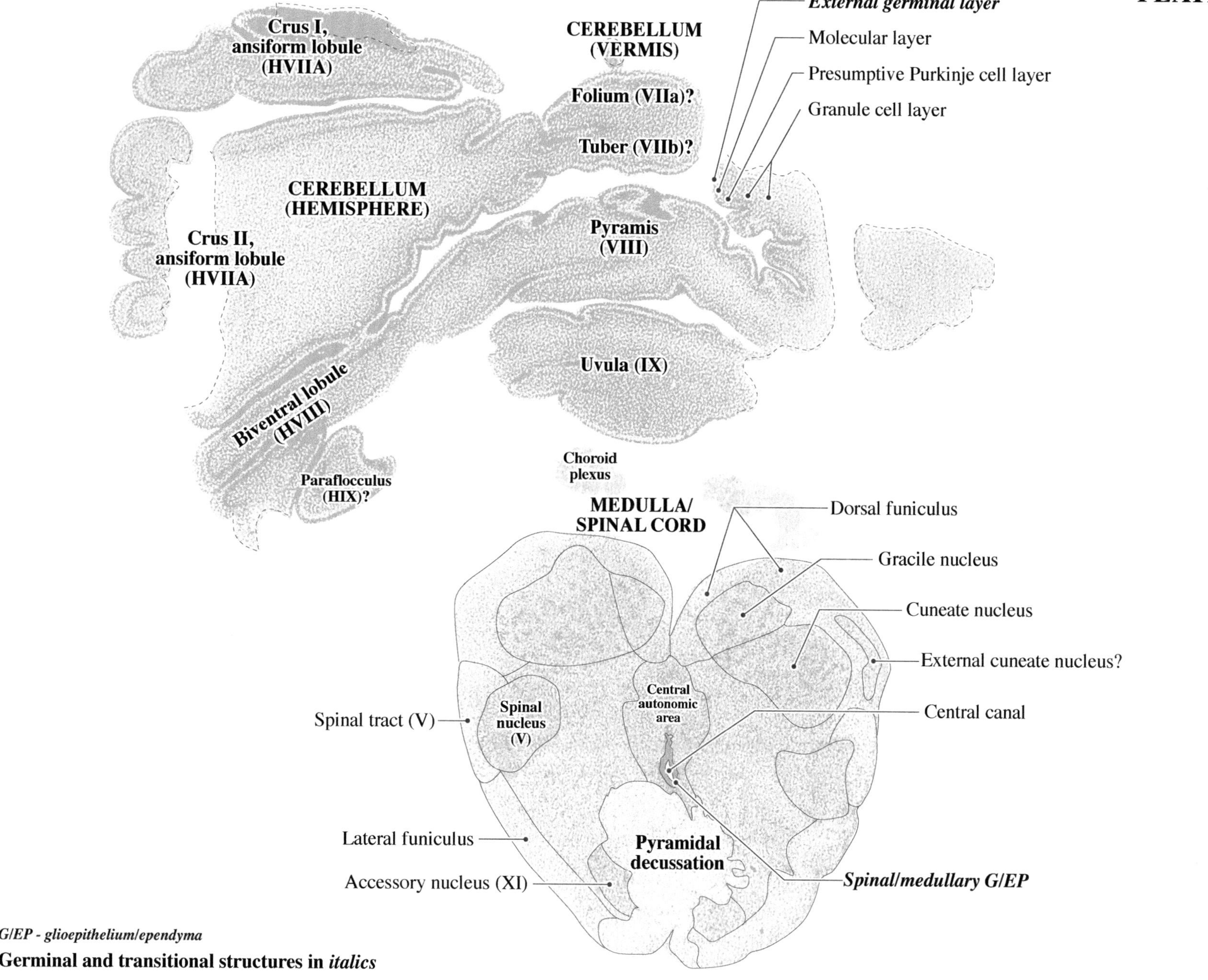

G/EP - glioepithelium/ependyma

Germinal and transitional structures in *italics*

PLATE 58A
CR 210 mm
GW 24, Y94-62
Frontal
Section 1201

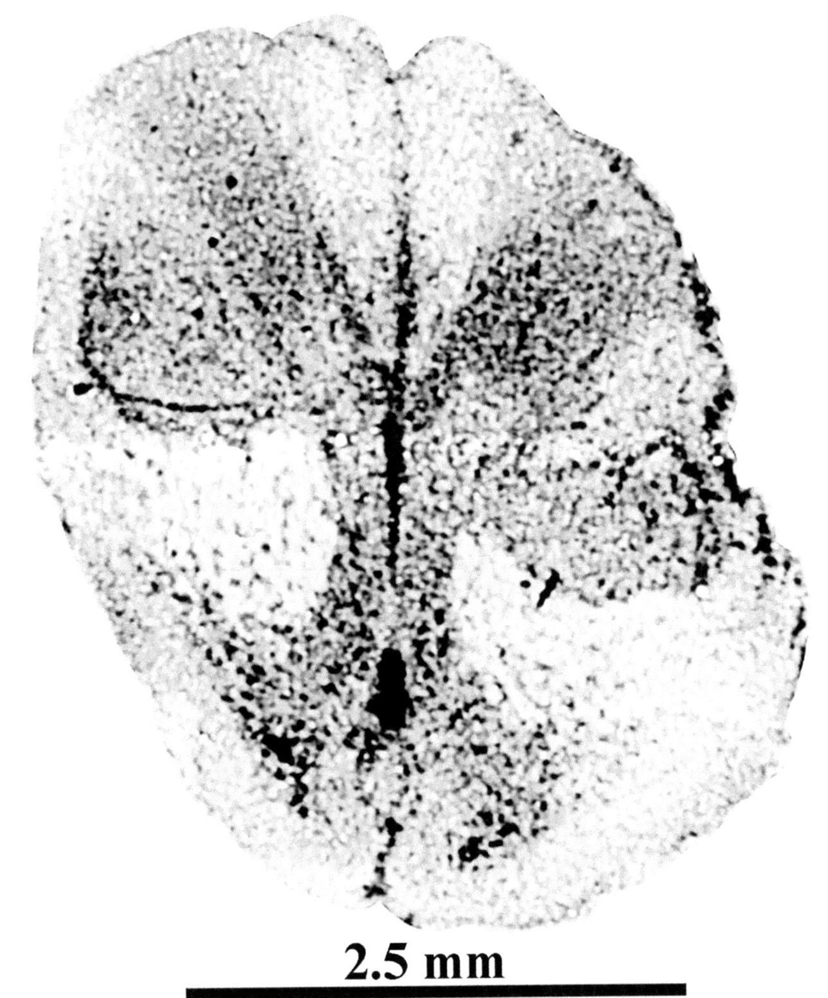

See the entire section in Plates 45A and B.

PLATE 58B

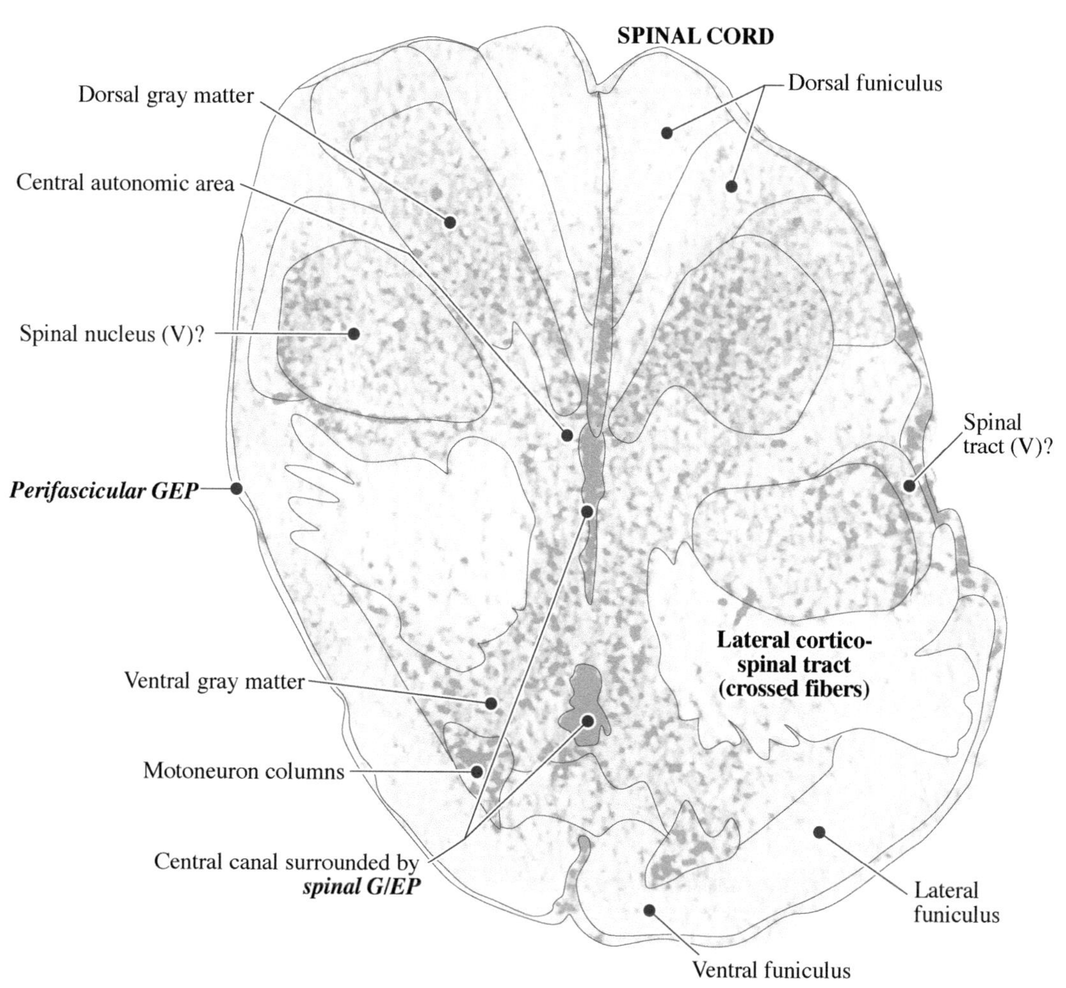

G/EP - glioepithelium/ependyma
GEP - glioepithelium
Germinal and transitional structures in *italics*

PLATE 59
CR 210 mm
GW 24, Y94-62
Frontal
Section 481
FRONTAL CORTEX

See the entire section in Plates 32A and B.

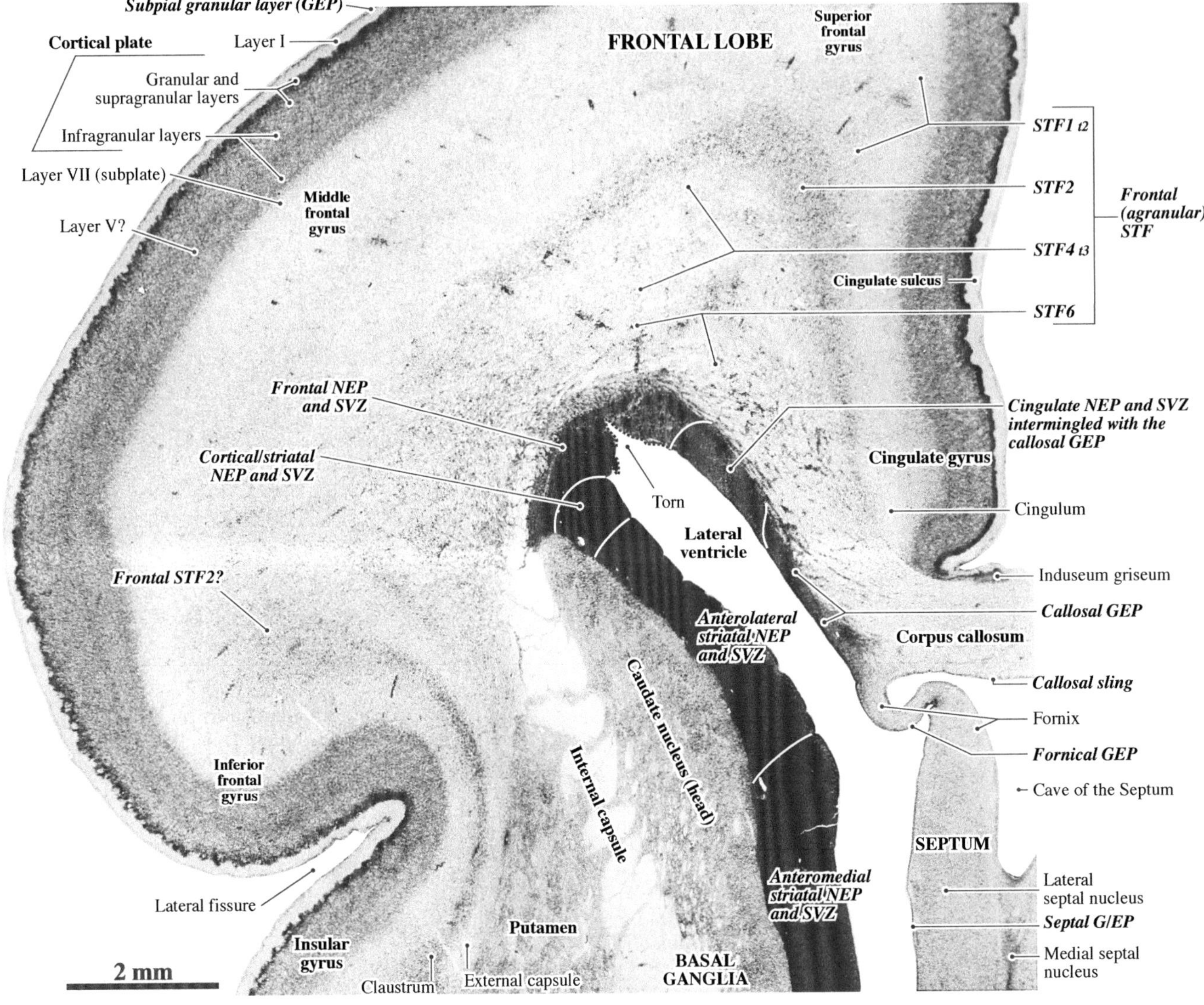

At this magnification, some lamination is visible in the agranular frontal cortical plate, where Layer IV will not be prominent in the mature brain. The dense superficial part is presumed to be composed of recently arrived neurons in Layers IV (granular), III, and II (supragranular) neurons. The less dense deep portion of the cortical plate is composed mainly of the earlier arriving neurons in the infragranular Layers V and VI; Layer V may be the slightly darker band. The thickest part of the developing cortex is the ***STF*** between Layer VII and the densely packed proliferating cells in the ***NEP*** and ***SVZ*** lining the ventricle. This area contains axons intermingled with neurons migrating to the cortical plate and glia destined to settle either in the cortex or the deep white matter. In this area of agranular frontal cortex, there are only 4 of the 6 layers in the ***STF***. ***STF2*** is the only layer that contains cells; ***STF4*** is in the ***t3*** phase when fibers are there after cells sprouting axons have moved up into the cortical plate. Fibers are plentiful throughout the second trimester in ***STF1 t2*** and ***STF6***.

The *white lines* indicate presumptive subdivisions (mosaicism) within the dense layer of proliferating cells lining the lateral ventricle. The *dashed lines* are torn parts of the frontal ***NEP*** and ***SVZ***. Thickness of the germinal matrix layer indicates the level of proliferative activity. The layer is very thin lining the septum and is presumed to be generating glia and the ependymal ventricular lining that will persist in the adult. In contrast, the proliferative layer thickens beneath the cortex and corpus callosum, and thickens more over the striatum (head of the caudate nucleus). The cortical/striatal ***NEP*** is a unique junctional area that may give rise to neurons destined for quasi-cortical structures such as the claustrum, endopiriform nucleus, and basolateral components of the amygdala. (***Abbreviations: G/EP-glioepithelium/ependyma, GEP,-glioepithelium, NEP-neuroepithelium, STF-stratified transitional field, SVZ-subventricular zone***)

PLATE 60
CR 210 mm
GW 24, Y94-62
Frontal
Section 821
PARACENTRAL/ TEMPORAL CORTEX

See the entire section in Plates 38A and B.

Lamination in the paracentral/temporal cortical plate shows a thicker band of dense granular and supragranular layers and thinner infragranular layers than in the agranular frontal lobe (*see* **Plate 59**); Layer IV may be the central dense band. This pattern is characteristic of granular (sensory) cortical areas. Generally, the granular cortex has a more elaborate array of cells and fibers in the ***STF*** than the agranular cortex in **Plate 59**. In both the temporal lobe and paracentral lobule, ***STF2*** is indistinct and has only sparse cells. In contrast, ***STF3*** is subdivided into a dense cellular band (***3a***), a fibrous core with vertically oriented cells (***3b***), and fragmentary groups of cells beneath the fibers (***3c***). The temporal cortex in this section does not have any STF layers deep to ***STF3c***. ***STF4*** in the paracentral lobule is subdivided into an upper late phase (***t3***) where fibers are devoid of interspersed cells, and a deep middle phase (***t2***) where migrating cells form striations between the fibers. ***STF5*** is present in only the most lateral part of the paracentral lobule; ***STF6*** is thick and filled with fibers that enter or exit the corpus callosum.

As in **Plate 59**, *white lines* indicate presumptive subdivisions (mosaicism) within the dense layers of proliferating cells bordering the lateral ventricle.

Abbreviations:
G/EP-glioepithelium/ependyma,
GEP,-glioepithelium,
NEP-neuroepithelium,
STF-stratified transitional field,
SVZ-subventricular zone

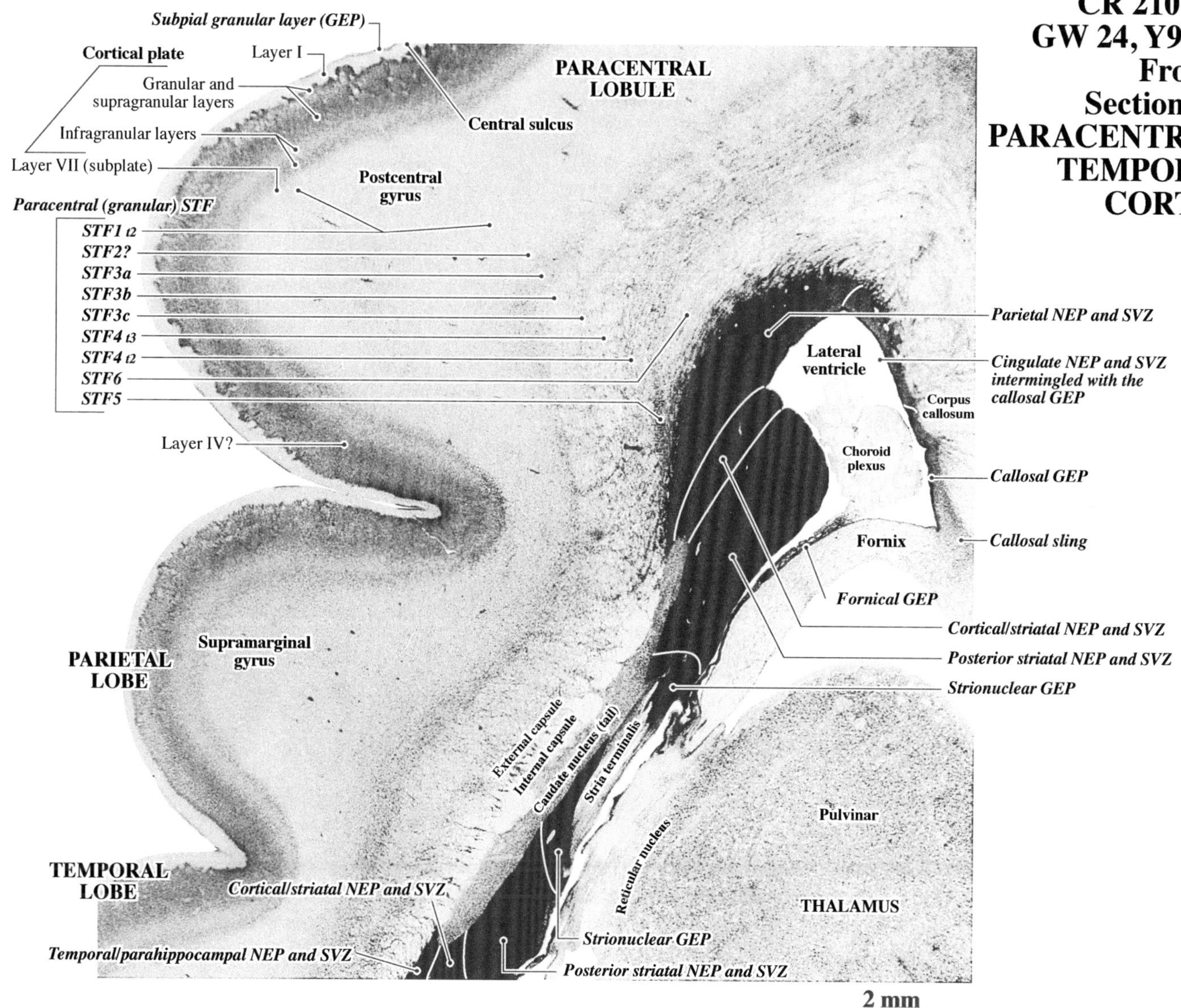

PLATE 61
CR 210 mm
GW 24, Y94-62
Frontal
Section 871
PARACENTRAL/ TEMPORAL CORTEX
See the entire Section 861 in Plates 39A and B.

Lamination in the pacentral and temporal cortical platse at this level shows many similarities with **Plate 60**. There are thick, dense granular and supragranular layers and thinner infragranular layers; Layer IV may be the central dense band. The ***STF*** shows many similarities with **Plate 60**. ***STF2*** is sparse compared to the dense array of cells in ***STF3a***. ***STF3b*** is beginning to have periodic vertical bands of cells between the fibers forming the "honeycomb matrix" that is most defined in the occipital cortex (*see* **Plates 43**, **44**, and **45**). ***STF3c*** is still not well defined from ***STF4***. ***STF4*** is in the late phase laterally (***t3***) and in the middle phase medially (***t2***). ***STF5*** is thin and may not be in the temporal cortex. ***STF6*** is absent in the temporal cortex in this section, but it is prominent in the parietal cortex where it blends with the corpus callosum.

As in **Plates 59** and **60**, the *white lines* indicate **presumptive subdivisions** (mosaicism) within the dense layers of proliferating cells bordering the lateral ventricle.

Abbreviations:
G/EP-glioepithelium/ependyma
GEP,-glioepithelium
NEP-neuroepithelium
STF-stratified transitional field
SVZ-subventricular zone

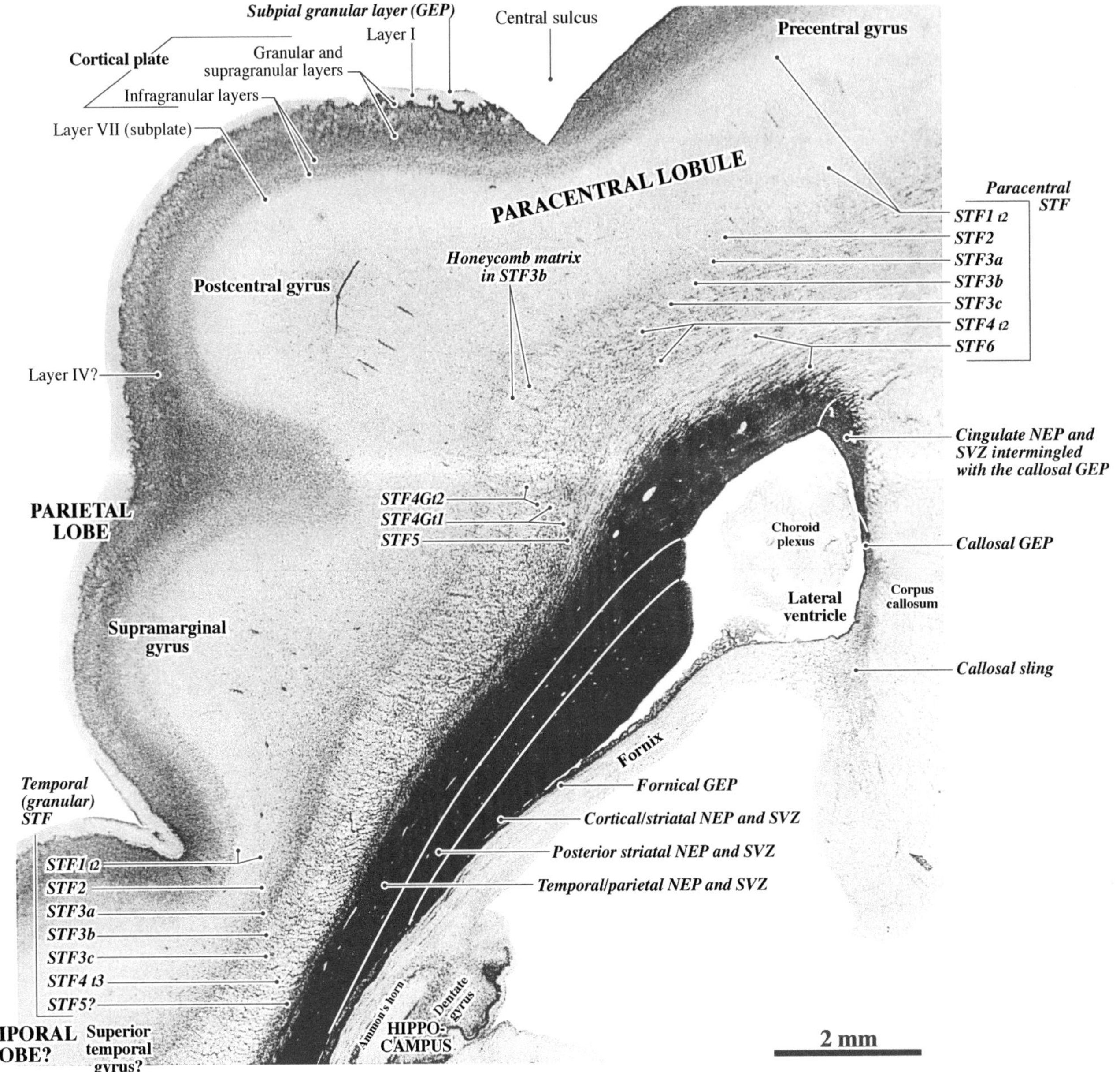

In order to show the full extent of the lateral and ventral cortex, the section orientation has been changed so that **dorsal** is at the right margin, **ventral** is left, **lateral** is top, and **medial** is bottom. As in **Plates 59** through **61**, the *white lines* in this section indicate **presumptive subdivisions** (mosaicism) within the dense layers of proliferating cells bordering the lateral ventricle. Lamination in the parietal/temporal cortical plate shows many similarities to **Plate 61** with dense granular and supragranular layers and sparse infragranular layers; Layer IV may be the central dense band.

STF2 is sparse; ***STF3a*** is dense; the periodic radially migrating cells in the honeycomb matrix are more evident in ***STF3b***; ***STF3c*** is fragmentary in the parietal cortex but continuous in the ventral temporal cortex. ***STF4*** is in the middle phase *(t2)* throughout its extent. ***STF5*** and ***STF6*** are more evident in the ventral than in the lateral temporal cortex.

(***Abbreviations: G/EP-glioepithelium/ependyma, GEP,-glioepithelium, NEP-neuroepithelium, STF-stratified transitional field, SVZ-subventricular zone***)

PLATE 62
CR 210 mm
GW 24, Y94-62
Frontal Section 911
TEMPORAL/PARIETTAL/ PARARACENTRAL CORTEX

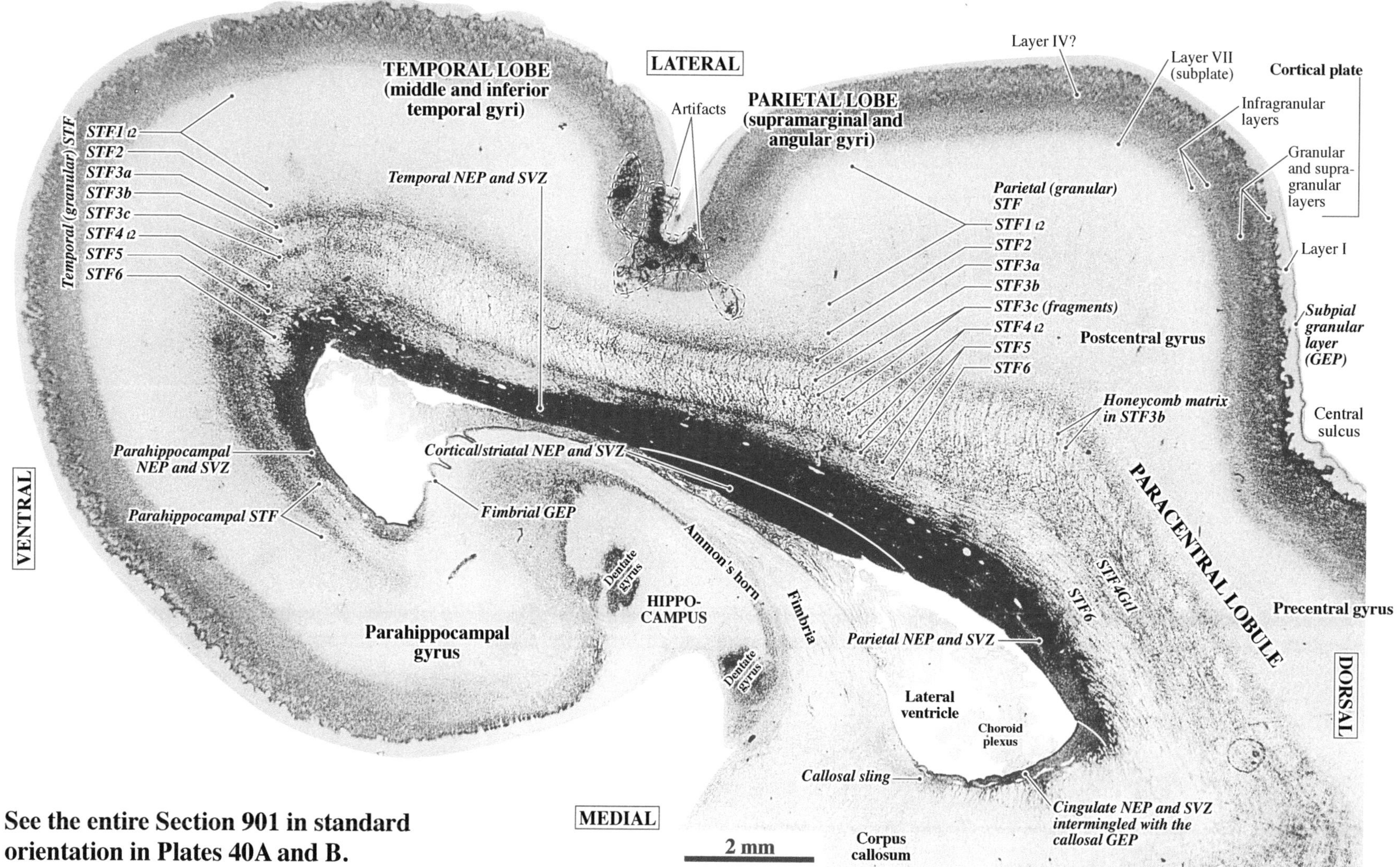

See the entire Section 901 in standard orientation in Plates 40A and B.

For Product Safety Concerns and Information please contact our EU representative GPSR@taylorandfrancis.com Taylor & Francis Verlag GmbH, Kaufingerstraße 24, 80331 München, Germany

T - #0292 - 160425 - C35 - 210/280/7 - PB - 9781032219394 - Gloss Lamination